Breaking New Ground

Agricultural Anthropology

Robert E. Rhoades

1984

INTERNATIONAL POTATO CENTER (CIP)

P.O. Box 5969 Lima - Peru. Cables: CIPAPA - Lima.
Telex: 25672 PE. Telephones: 366920 - 354354.

CIP is a nonprofit, autonomous scientific institution established by agreement with the Government of Peru for developing and disseminating knowledge for greater use of the potato as a basic food. International donors funding technical assistance in agriculture finance the Center.

Correct citation:

Rhoades, Robert E. 1984. Breaking New Ground: Agricultural Anthropology. Lima, International Potato Center. 1984. 84 pp.

XII-SS-E-10-0-4500

December 1984

CONTENTS

FOREWORD

In the past the relationship between agricultural and social scientists has been an uneasy one. Social scientists were often seen as outside critics who leveled sharp and unsympathetic attacks against agricultural research and development, especially in Third World countries. Anthropologists in particular seemed aloof, arbitrary, and arrogant toward biological science research.

While anthropologists viewed themselves as defenders of traditional agriculture against the negative effects of modernization, individuals working to improve food production saw the social scientists' relationship to rural populations as unbalanced. Anthropologists came, lived in villages for a year or more, and went home to publish books and articles in their language—read mainly by other anthropologists. While many an anthropologist rose to full professorship on data drawn from farm families, little or nothing remained behind to improve the lives of their "informants."

The sad irony is that agricultural development could have benefitted from anthropology. Insights gained from close contact with the everyday lives of farmers and ethnographic methods to help pinpoint areas needing technological improvement are but two examples of how anthropologists can have a positive input.

At the International Potato Center, an atmosphere has been developed where anthropologists are able to make a positive contribution in solving food production problems. Our philosophy encourages the anthropologist to work on interdisciplinary teams from the very beginning of a project and not as an after-the-fact critic.

Indications are that the mood in anthropology has now shifted toward a positive interest in agricultural research and development. This monograph illustrates clearly and succinctly this newly evolving kind of social science input. To the extent that it succeeds, agricultural research strategies to improve food production will be richer and more effective. Anthropologists, in turn, will contribute directly to worldwide efforts to intensify food production and therefore to the well-being of the people they study.

R.L. SAWYER

Director General
International Potato Center
Lima, Peru

PREFACE

Toward the end of the Great Depression, two young American social scientists — Robert Redfield and W. Lloyd Warner — published a now-forgotten article "Cultural Anthropology and Modern Agriculture" (Redfield and Warner 1940). In outlining anthropology's potential role for understanding and solving practical problems in agriculture, they argued "the science of Man" offered a unique perspective on how farming people adapt to their social and physical environments. Anthropology's explanatory theories and rural-based field methods seemed especially promising for planning and executing agricultural programs.

For reasons unknown, however, Redfield and Warner (1940:992) explicity wondered if the practical application of anthropology to agriculture would ever occur. In fact, time has almost proven their misgivings correct. Over the four decades since the article appeared, the paths of anthropologists and agricultural scientists rarely crossed, a most surprising circumstance since anthropologists have dealt more directly and intimately with farming peoples than any other group of social or biological scientists. Emerging only now is the shadow of an "agricultural anthropology" that aims to move anthropology into the mainstream of agricultural research and development.

In this monograph, I wish to explore anthropology's past, present and future in humankind's most essential enterprise. Three key questions will be discussed.

1. Why, in the past, did anthropology fail to develop an applied subfield dedicated to agriculture?
2. Where, in the present, is anthropology in relation to agricultural research and development?
3. How, in the future, should anthropology proceed if agriculture is to benefit from anthropological understanding?

To answer these questions, I will draw on my own experiences over the past 20 years in both anthropology and technical agricultural development. Coming from a farm background and agricultural training I entered the development work as an extensionist in Nepal in the early 1960s, well before the onset of the "Green Revolution." In the late 1960s, I was in the Philippines studying agricultural education and conducting research on the diffusion of "miracle rice" which was changing Asia's agricultural landscape. Finally, in the late 1970s after 10 years in academic anthropology, I became one of the first anthropologists to work as a senior scientist in an International Agricultural Research Center.*

---

* For background information on the system of International Agricultural Research Center networks and the Consultative Group on International Agricultural Research see CGIAR, 1980.

In each phase of my own involvement I looked upon planned agricultural change with different emotions, ranging from naive optimism to intellectual pessimism, and finally, to a more seasoned realization of what is possible, needed, and desired if our hungry planet is to survive. Throughout this process, I have become convinced that anthropology can and should play a positive role in agricultural research and development. This monograph, hopefully, will help to better understand how far we have come and how far we need to go in this new field that we have begun to break.

Robert E. Rhoades
Lima, Peru
July 26, 1984

Acknowledgments

Many individuals have contributed ideas and comments for this monograph. Prominent among them were Robert Werge, who pioneered and created a future for anthropology at CIP, and Robert Booth, post-harvest technologist, who encouraged and debated this new anthropology. A special thanks to Ralph K. Davidson and Gelia Castillo who so patiently encouraged me in my research. I also gratefully acknowledge the useful comments on earlier versions of this paper by Susan Almy, Guy Camus, Billie Dewalt, Curtis Farrar, Douglas Horton, Angelique Haugerud, Jeffrey Jones, Vera K. Niñez, Jorge Recharte, Vernon Ruttan, Greg Scott, Robert Tripp, Stephen Thompson, Helga Vierich, and William F. Whyte.

I. ANTHROPOLOGY and AGRICULTURE: THE HISTORICAL INTERFACE

Since founding of the discipline until the present, anthropologists have studied virtually every manifestation of agriculture ranging from archeological investigation of the orgins of agriculture down to ethnographies of modern industrial farming, Flannery 1965; Reed 1977; Bennett 1969, Barlett 1980). This emphasis has not been misplaced. In terms of sheer numbers, most of humankind are still tillers of the soil or are directly engaged in other forms of food production such as herding or fishing. Today, as always, the fundamental business of nations —including most industrial ones— is agriculture. Basic food production remains the primary human activity and, with increasing population pressure against finite natural resources, people will continue to seek ways to improve and intensify that production (Rhoades and Rhoades, 1980).

How is it possible, then, that anthropology is simultaneously a discipline that has dealt so directly and intimately with the world's rural people while having virtually no direct involvement in, or impact on, planned agricultural change? The answer lies in part in the history of anthropology and its institutional connections.

The Past: Lost Opportunities

Anthropology's rural bias and concern with contemporary Third World peoples are tied directly to the very foundations of the discipline. In the mid-19th century when formal social science disciplines began to develop in Western universities, the desire of each new discipline was to mark off territory in the study of European institutions and the Great Civilizations. History thus became the study of Western man, sociology the study of Western institutions, and behavior and economics the study of European commercial and capitalist conditions.

What remained for anthropology, the youngest discipline? Anthropology has often been referred to as the "slum child of the sciences," accepting subjects rejected by the established social science disciplines (Beals 1973). The focus thus became the mass of humanity colonized by Europe, called at that time "tribal," "uncivilized," "savage," "native," and "exotic." Living largely in the marginal areas of Africa, Asia, North America, and Latin America, these same peoples today are called in development language "clients" or "target populations" for agricultural development programs.

Anthropologists studied these populations, often composed of small groups or communities, in their physical and cultural entirety. Thus, early anthropologists

1

were simultaneously sociologists, economists, historians, biologists, linguists, and technologists of colonized peoples. With total societies to consider, anthropologists attempted to duplicate in almost every aspect what specialized disciplines were doing for different segments of European culture. Anthropology's contemporary impatience with the application of European models of psychological or economic behavior to subsistence agricultural populations can be traced to early field experiences with human systems radically different than those of European origin (see Wolf 1964 for one of the best histories of anthropology).

The discipline served as an important information link between the colonized, often tribal or peasant populations, and colonial or neo-colonial government (see Foster 1969: 181-217 for an early history of applied anthropology).[1] This colonial role primarily centered on the description of traditional ways and impact of European contact. Public health, education, and community development loomed as important early applied fields for anthropology.

Except for a temporary involvement with the Soil Conservation Service of U.S. Department of Agriculture in the 1930s (Foster 1969: 202), however, anthropologists rarely touched base with applied agricultural researchers.[2] A notable exception was sociologist Charles Loomis (1943), one of the founders of the Society for Applied Anthropology, who spent half a year at Tingo Maria, Peru, analyzing problems of establishing an agricultural extension service. Loomis also worked in Costa Rica in the late 1940s and early 1950s where he helped establish the social science department in Turrialba which ultimately evolved into the Institute Interamericano de Ciencias Agricolas (IICA) of today.[3] The Peruvian Vicos Project, although not exclusively agricultural, subsequently had an impact on planned agriculture and policy in Peru (Holmberg, et al. 1962, 1965).[4]

Perhaps the most perceptive early work illustrating the uses of anthropology for practical agriculture is Pierre De Schlippe's (1956) *Shifting Cultivation in Africa.* De Schlippe, a practicing agronomist, discovered in the course of his work in Africa the great benefit of anthropological methods and theory. In what may be the first major publication advocating agro-anthropology he set out to undertake research on the "borderline of agronomy and anthropology." His concerns (1956:XVI), so well stated 30 years ago, still go largely unheeded:

---

[1] In the United States, anthropology's main applied involvement was in the Bureau of Indian Affairs. The establishment of the Bureau of American Ethnology in 1879 gave anthropology a stronger legitimate applied role within the U.S. government but no projects were concerned with farm problems *per se*. Parallel, to these developments, however, a subdiscipline of economics—farm management—was developing its own governmental bureaucracy in the United States which evolved into the Bureau of Agricultural Economics.

[2] Burleigh Gardner, Solon Kimball, and John Province worked for the Soil Conservation Service. Carol Taylor, a rural sociologist, employed Horace Miner, Oscar Lewis, and Walter Goldschmidt (1947, 1978) to do community studies in several states for the Department of Agriculture (Werge 1977:6).

[3] I am grateful to Jeffrey Jones for bringing this information to my attention.

[4] For example, Mario Vasquez—a key figure in the Vicos project—served as Director General of Agrarian Reform and Rural Settlement from 1973-76.

> Neither the agricultural research station nor the field anthropologist alone can give us the necessary understanding of agricultural practice in the humid tropics. The crucial problem, perhaps not even of Africa alone but of humanity as a whole, lies in the contact zone between man and his environment, between inhabitant and habitat, and therefore between two fields of research which have not yet undergone the necessary co-ordination.

Although important, these early efforts led to no lasting contribution by anthropologists to agricultural programs or the formation of an applied subdiscipline dedicated to agricultural problems. Edward Montgomery and John W. Bennett (1979) place part of the blame on the academic anthropology establishment for turning its attention away from food and nutrition matters in the 1950s and 1960s. The 1940s had witnessed considerable anthropological involvement in food and nutrition studies in the United States. Active in these efforts were outstanding professionals such as Margaret Mead and Robert Redfield. Developments in the universities in the post-war period, however, set anthropology upon a "return voyage to tribal ethnology and theoretical interests" and away from applied anthropology.[5]

They explain the consequences

> ... professional rewards have been given mainly to those anthropologists who have excelled in the traditional fields centering on the study of tribal and peasant humanity. Indeed, few examples can be cited of individuals preeminent as specialists in applied anthropology and contemporary complex society who have been given entry to the inner circle of the profession. Further, in the United this has meant that the American Anthropological Association has not featured applied work as a major subdivision of the general discipline, nor has it invested significantly in the publication of research with contemporary themes (Montgomery and Bennett 1977:127).

William Whyte (1984) further argues that in part anthropology's aloofness toward planned agriculture derived from the anthropological worldview in the 1950s and 1960s:

> I believe there was a tendency in those years for many social anthropologists to look upon culture as if it were cast in concrete rather than thinking of culture in terms of a framework developed by the people to handle the problem of living in their particular environment and therefore flexible to admit a modification when they could see that change would be advantageous. In that era, I was inclined to attribute the failure of plant and animal scientists to include anthropologists and sociologists in their programs to resistance by biological scientists who wanted to maintain their own monopoly in the field. Now, although I do not doubt that such resistance was important, it seems to me that the prevailing orientation of many anthropologists in that era was self-defeating, insofar as their gaining partnership in these programs.

Anthropology's institutional location in the university structure has also been detrimental to involvement in planned agriculture. In the United States, most land grant agricultural colleges do not have anthropology programs. Even if one

---

[5] In commenting on an earlier version of this paper, economist Douglas Horton noted: "How many anthropologists wanted to participate in agricultural projects? I have always been struck by my anthropology colleagues' concern with academic independence, intellectual honesty and other pretensions of purity, always feeling apart from and superior to the technicians and agronomists. Was this the source of the isolation problem, or the result (sour grapes)? Of course, Bill Whyte, Holmberg, Barnett, were early allies. And they have been accused by their more purist colleagues for selling out."

3

exists, it is generally located in liberal arts, not in science colleges. This structural position left anthropology isolated from planned agriculture during the early years of the post-war expansion of United States involvement in agricultural programs both at home and overseas. Throughout this key post-war period, few anthropologists entered development. At most, ethnographers were known for helping expatriates adjust to "cultural shock" or to interpret the "silent language" and "hidden dimension" of foreigners' behavior but not for making a substantial contribution to planned change (Hall 1961; 1966).

It is also logical that international technical agricultural program leaders of the early USDA and "Point Four" programs would turn to their closest social science colleagues, agricultural economists, for input on the human component of agricultural development. Agricultural economists, however, did not gain their ground simply because of an inherited inside track. Economics as a profession has a "client relationship with society" (Thurow 1977:80). Anthropologists, on the other hand, were known to be arrogantly aloof, refusing often on moralistic grounds to apply their expertise to practical agriculture despite the wealth of anthropological knowledge available on primitive and peasant agriculture. Anthropologists Robert Netting (1974) has further argued that agriculture was simply considered . . .

> too basic for the ambitious new science of man, and the suspicion lingers that it is intellectually "infra dig." The supposed simplicity, concreteness, and lack of system in most non-western farming techniques did not attract minds stimulated by the complex, abstract order of kinship terminology, descent groups or ritual patterns.

It might have been, however, that the complexity of tropical agriculture puzzled technically naive anthropologists. A few ethnobotanists, cultural ecologists, and archeologists conducted important studies in the 1950s and 1960s revealing the finely adapted nature of agricultural and horticultural societies to social and environmental conditions. Some, such as Conklin's (1954, 1957) detailed study of Hanunoo agriculture, and Geertz's (1963) *Agricultural Involution,* have been widely read and cited by agricultural scientists. Anthropology's fundamental concern with basic food production, even in the origins of agriculture, is evidenced in any modern anthropology textbook (Harris, 1971). Twenty percent of the articles in *Plant Agriculture,* a collection of readings from *Scientific American* magazine articles 1950 to 1969, were written by anthropologists and archeologists. In fact, the rural element is so deeply ingrained in anthropology that the term "rural anthropology" would strike many people as humorously redundant while "urban" anthropology is a legitimate subfield. Missing, however, was anthropology's involvement in planned agricultural change.

The Present: International Agricultural Research

The watershed period marking the significant inclusion of social science into international agricultural research must be traced to pioneering efforts in the early 1960s at the Philippines' International Rice Research Institute (IRRI).

Vernon Ruttan (1982: 308-309), the first economist to work in the international center system, recorded his early experiences that are not unlike those of anthropologists now entering agricultural research:

4

When I arrived at IRRI, I was shown to an office in the very attractive new institute complex. The office was conveniently located near the library. It had a brass plate in the door with the label *Agricultural Economics*. In the weeks that followed, however, neither the director nor the associate director of IRRI conveyed to me a very clear idea of why they needed an agricultural economist or what contribution they expected from the economics unit at IRRI.

Agricultural economics went on from that point to make its mark on the international research center developing both the "constraints research approach" and later promoting "Farming Systems Research." Today, all but one international center in the Consultative Group on International Agricultural Research (CGIAR) have economics programs. At CIMMYT, the international wheat and maize center headquartered in Mexico, economics is the third major research program along with technical research on the two crops.

Anthropology's chance did not come until more than a decade later, in the mid-1970s. It is doubtful at this time that an international agriculture research center would have hired an anthropologist with core funds. Few administrators in agricultural research management had a clear idea what anthropology was or how it was relevant to their technical programs. Those few who did were economists who were concerned, as some still are today, with holding the limited turf they had so arduously gained. Sharing ground with anthropologists was an uncertain and threatening thought. Anthropologists had a reputation as critics of the "Green Revolution" (Ryan 1979:120). It was feared that the entire house economics had built might come tumbling down through a negative reaction by management, generally made up of biological scientists who sometimes lumped both anthropologists and economics into the category "social scientists," considered more alike than different.

The Rockefeller Foundation finally paved the way for the inclusion of the "non-economic social science perspective" by establishing and funding in 1974 its "Social Science Research Fellowship in Agricultural and Rural Development." The purpose was *not* to permanently place social scientists in the international centers but to give recent North American PhDs experience in agricultural development. Whether by accident or design, the majority (21 of 33) of Rockefeller post-doctorates assigned to posts between 1975 and 1984 have been anthropologists.

In reflecting on the early years when the Rockefeller program was just getting underway, Susan Almy — then with the Foundation — writes:

The biggest objection against anthropologists was that they sat in a single village whereas the centers were mandated to create knowledge useful at an international level. The RF fellowship program was begun in 1974 but only opened up to non-economists in 1975. The emphasis on anthropologists was entirely due to the greater response by them to the opportunity. I was very careful to advertise and write all the major agricultural economics, economics, geography, rural sociology as well as anthropology departments and journals. Among the other disciplines, good students tended to want to stay home after the degree, whereas the tide had just turned in anthropology and many students wanted nonacademic experience (Susan Almy 1984).

A parallel development to the Rockefeller Foundation efforts was the formation in late 1976 of an American-based group known as the Anthropological Study Group on Agrarian Systems (ASGAS) or, after the name of its bulletin, the *Culture and Agriculture* Group. This organization has been instrumental in giving visibility to agricultural research among anthropologists (Barlett 1980:546). A similar informal association of anthropologists formed in Lima, Peru, in 1979 but with a more international focus than the US-based ASGAS group. These late 1970s developments testify to the resurgence of interest in agriculture and food among anthropologists, an interest — along with Farming Systems Research — which has gained considerable momentum in the 1980s.

II. The Case of the International Potato Center

Not surprisingly the International Potato Center (CIP) was the first to receive a Rockefeller Foundation social science research fellow and has subsequently gone on to utilize more anthropologists than any other world agricultural research organization. First, CIP is located in Peru where anthropology is as strong in rural development, if not stronger, than economics. Second, the economists on the CIP staff were not antagonistic to "non-economists"[6] and had formed strong personal and professional links with anthropologists. This, in part, may be due to the marked ecological and human landscape of the Central Andes that has allowed a fertile interchange between the two disciplines. Finally, CIP—a new research center in the early 1970s—was seeking creative approaches to development. While it may not have been initially clear how anthropologists would concretely contribute to CIP's goals, Director General Richard Sawyer viewed anthropology and sociology as new and potentially useful disciplines.

Since 1975, 14 anthropologists and two sociologists have worked in some capacity or conducted research in direct association with CIP. The majority have come to the Center with their own funding or held temporary appointments. Two, both Rockefeller researchers, have been hired on the permanent senior staff although not simultaneously. In 1984, two continuing positions open to anthropologists or sociologists were added to the department. The CIP experience thus offers a unique experiment to illustrate the potential strengths and weaknesses of anthropology in agricultural research.

Introduction of Development Anthropology and Sociology

The future of any new discipline in an agricultural research center rests in part on its early experiences with individuals representing that discipline. CIP was fortunate in that its first "non-economists" were individuals who understood and related well to agricultural scientists.

---

[6]I dislike using the term "non-economists." It is, however, a linguistic category widely used in agricultural research. Unfortunately, it masks over the economic (not necessarily neoclassical) orientation of many anthropologists and lumps together sociologists and anthropologists. In many respects, sociology and anthropology are more distinct than anthropology and economics or sociology and economics.

Among these were Clyde Eastman, a rural sociologist from New Mexico State University. Not only did Clyde look, dress and talk like an "aggie" (tremendous symbolic assets in agricultural development), he set down some basic principles which CIP economists and anthropologists have followed ever since. The first was plain, straight language in presenting the social science point of view. Although the institutionally more powerful biological scientists are notorious for using jargon, they tend to be intolerant of social science jargon.

Eastman's second principle was to keep all written reports short and to the point (see Appendix of this report for three applied anthropological papers). In Clyde's 6 months sabbatical with the Center he generated a series of brief papers with such eye-catching titles as "The Cold Hard Realities of Agricultural Development" and "Should Peru Promote Potato Production?" These were widely read and debated by biological scientists in the Center, as it struggled to define its goals in the early years. Without a doubt, Eastman helped established a favorable impression of "non-economists" in the minds of the administration. Such pioneering efforts should never be overlooked in the varied fates that sociology and anthropology have had in other international agricultural centers.[7]

Another development anthropologist-sociologist who established contact early in the life of the Center and made a positive impression was William F. Whyte of Cornell University. Although not officially attached to the Center he offered constructive ideas on institutional building and mechanisms of technology development and transfer. Later CIP called on his expertise as consultant in the analysis of potato seed systems in Colombia, a work which influenced future socioeconomic research on seed. His practice of "applied" instead of "academic" anthropology helped correct the false image that anthropologists are only seekers of information about the quaint and exotic. And, like Eastman, Whyte's reports were succinctly and nonjargonistically written (Whyte, 1977).

Ecological Anthropology and Appropriate Methods

Following recommendations of a 1976 planning conference, the young and relatively inexperienced Social Science group began to acquire expertise by concentrating research on a specific Peruvian potato producing region (Mantaro Valley) where it developed methodologies and built closer links to biological research activities. This social science program was funded by IDRC-Canada and subsequently called "the Mantaro Valley Project" (Horton, 1984).

The leader of the Social Science Unit at this time was Douglas Horton, an economist, whose personal ties with anthropologists were perhaps stronger than with economists. This was due in part to the long standing involvement in Peru of Cornell University (from which Horton obtained his PhD). When it came time to gain an initial overview of the selected region—the Mantaro Valley—and where CIP has its major Peruvian research station, it was only logical that Horton would turn to an individual and friend who knew the valley well. This was Enrique Mayer, a Peruvian anthropologist, born and reared in the Mantaro Valley

---

[7]To my knowledge, six of the 13 CGIAR international centers have utilized anthropologists. At least three have decided that anthropology was not worth contining on core funding. CIP and CIMMYT are the only Centers to make a solid long-term commitment to anthropology.

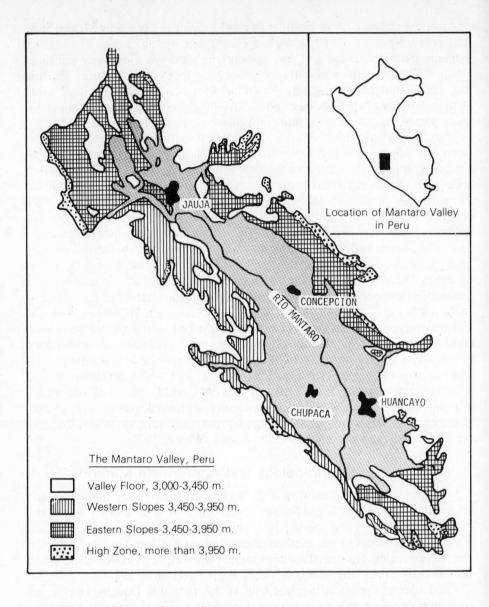

Location of Mantaro Valley
in Peru

The Mantaro Valley, Peru

□ Valley Floor, 3,000-3,450 m.

▥ Western Slopes 3,450-3,950 m.

▦ Eastern Slopes 3,450-3,950 m.

▨ High Zone, more than 3,950 m.

A CIP ecological anthropologist, using aerial photographs and ethnographic ground truth techniques, defined within a few weeks the major agroecological zones, types of producers, and cropping patterns of the Mantaro Valley. Based on this information, an agroeconomic team proceeded with on-farm trials with representative farmers (Mayer 1979).

and who, like Horton, formed part of the Cornell "Peruvianist" group that post-dated the often bitter feuds of the Vicos project.[8]

The task assigned Mayer in the "Mantaro Valley Project" was to conduct an initial anthropological overview of the valley, define its major zones of production, and delineate types of producers found in each zone. Given Mayer's long-term knowledge of the valley and rapid survey methods, the task was completed in 2 months on a small budget of under US$ 2,000. The resulting publication with its land-use map has subsequently come to be a widely read anthropological study of an Andean region as well as an inspiration to social scientists in terms of what can be done so rapidly and expertly (Mayer, 1979).

Mayer utilized the theoretical perspective of ecological anthropology to define the major agroecological zones and land use patterns within the Mantaro Valley. His use of aerial photos, government data, and ethnographic "ground truth" techniques are excellent examples of how anthropology can be a powerful discipline to help focus agricultural research projects. With this information in hand, the newly constituted CIP agroeconomic team composed of economists and agronomists could proceed with planning and executing on-farm trials using known technologies as well as technology being generated on the CIP experimental station (Horton, 1984).

In addition to practical utilization for planning on-farm trials, the Mayer effort also set the foundation for illustrating how informal survey methods of anthropologists could be used for rapidly and inexpensively gaining an overview of agricultural land-use and cropping patterns in a region. This work went on to form the basis of future CIP methodological studies on informal or rapid rural surveys appropriate for developing countries (see Rhoades, 1982a).

Ethnobotanical Research

A major reason for CIP's existence is the collection and maintenance of a world germplasm pool of wild and cultivated native South American potatoes. This germplasm "bank" contains natural resistances that can be utilized by breeders in improving potato varieties for developing countries.

The complex folk nomenclature of native potatoes used by Andean farmers has long fascinated both anthropological and biological scientists (La Barre 1947; Hawkes 1947). However, how or why this information might be useful to a technologically oriented center such as CIP was never made clear. Ethnobotanical studies conducted in 1977-78 by anthropologists Stephen Brush and Heath Carney provided basic information on farmer selection of varieties useful to the Center's efforts at collection and maintenance of a world germplasm pool for utilization in developing countries. Their research revealed that farmers use a four-level system of classification, integrating wild, semi-domesticated, and domesticated species. Instead of a chaotic and random system, as is often assumed by outsiders, the researcher's data revealed a system of farmer classification, selection and use of native varieties as logical as the modified Linnean system used by biological scientists (Brush et al. 1981). The realization of a complex native folk taxonomy prompted biological scientists to pay closer attention to the native classification systems and nomenclature for cataloging germplasm collection.

---

[8]William F. Whyte was adviser to both Horton and Mayer at Cornell.

The World Potato Collection of the International Potato Center is made up of all known existing wild and domesticated potato species. More than 10,000 possibly different native varieties still grown by Andean farmers have been collected and preserved. Anthropologists have contributed to potato improvement through studies of ethnobotany, consumption, and farmer strategies in germplasm use.

Other ethnobotanical data were useful for design of on-farm experiments. For example, Steve Brush demonstrated that mixed plots planted in native varieties are valued for their home consumption and culinary qualities while improved varieties are intended for market or exchange and planted homogeneously. For on-farm research it is important to understand this simple difference in farmer strategies. Otherwise, comparisons of home consumption plots with recommended, commercially-oriented trial plots may not make sense within the farmer's dual strategies.

Nutritional Anthropology and Consumption Research

CIP contracted an anthropology post-doctoral researcher to study potato consumption and nutrition. Susan Poats' research focused on the role of potatoes in the human diet and the actual and potential impact of potatoes on nutrition in developing countries (Poats 1983). This research project has helped to better understand the preferences of consumers, the ultimate clients of the International Potato Center. It also brought a degree of cultural relativism to the Center's general thinking by emphasizing the important role of consumer preferences for color, taste, shape, and cooking quality in the selection of varieties. This research also helped dispell several myths about potato consumption and provided policy makers with a more solid basis for appraising the value of the potato as a food crop in their countries.

Specifically, CIP breeders are now aware that considerably more variation in color, shape, and size of potatoes may be acceptable in developing countries than in Euro-American countries whose markets demand uniformity in tuber size and color. The study also emphasized the nutritional importance to developing coun-

Transect of native varieties of potato in a field near Chinchero, Peru, Cuzco Dept., alt. 3820 meters.
(Brush etal. 1982)

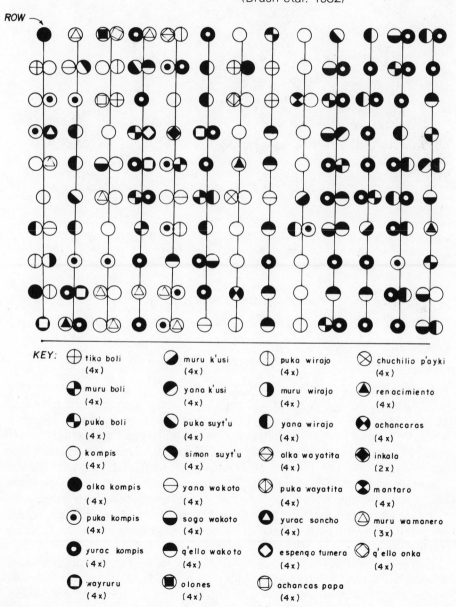

KEY:

⊕ tika boli (4x)	◐ muru k'usi (4x)	◖ puka wirajo (4x)	⊗ chuchilio p'ayki (4x)
◕ muru boli (4x)	◢ yana k'usi (4x)	◑ muru wirajo (4x)	▲ renacimiento (4x)
⊕ puka boli (4x)	◣ puka suyt'u (4x)	◖ yana wirajo (4x)	⊗ achancaras (4x)
○ kompis (4x)	◣ simon suyt'u (4x)	⊖ alka wayatita (4x)	◈ inkala (2x)
● alka kompis (4x)	⊖ yana wakoto (4x)	⊕ puka wayatita (4x)	✕ mantaro (4x)
⊙ puka kompis (4x)	⊖ sogo wakoto (4x)	△ yurac soncho (4x)	△ muru wamanero (3x)
● yurac kompis (4x)	◗ q'ello wakoto (4x)	◇ espenqo tumera (4x)	◇ q'ello onka (4x)
▢ wayruru (4x)	◉ olones (4x)	▢ achancas papa (4x)	

11

An anthropological follow-up of potato germplasm use in Nepal. The farmers are helping the investigator identify positive and negative characteristics of the varieties they grow.

try consumers of dry matter content, a preference now being taken into consideration in selection of germplasm materials.

Comparative Anthropology Applied to Agricultural Research:
Peruvian Farming Systems in Worldwide Perspective

Anthropology has also made a contribution to CIP's program with its comparative methodology. A major problem in international agricultural research is extrapolation; that is, are results in one region relevant to other similar regions in areas of the world. The International Potato Center has long justified its programs in terms of conducting research in different agroecological zones of Peru presumed to be representative of other areas of the world. CIP conducts research in four distinct Peruvian regions: La Molina and Cañete on the arid coast; Mantaro Valley, in the highlands; San Ramon, in the humid, high jungle; and Yurimaguas, in the low Amazon Basin. These zones offer excellent natural laboratories for biological and agronomic research on the potato under different conditions.

However, this research has been mainly conducted on experiment stations and related to variables specifically of interest to breeders, virologists, and agronomists, (temperature, soils, and precipitation). Relevancy of on-station research for on-farm conditions and existing farming systems in surrounding communities was not clear.

The Social Science Department up to 1979 had concentrated its efforts only on CIP's highland research site, the Mantaro Valley (Horton, 1984). My first research assignment, therefore, was to describe, as Enrique Mayer had done in the Mantaro Valley, the farming systems in the communities surrounding CIP's experiment sites on the coast, high jungle, and low jungle.

Peru presents to the anthropologist a superb natural laboratory for studying distinct farming systems and how a crop like the potato fits, or might potentially fit, into these diverse farming systems. With the help of two Peruvian anthropology students, the farming systems of Cañete (arid coast), San Ramon (humid, hill zone), and Yurimaguas (the low jungle) were studied using informal survey methods (Recharte 1981, Bidegaray 1981). These sites were to be compared to the highlands to discover similarities and differences between CIP's Mantaro Valley efforts and research being conducted in the lowland zones. These data, in turn, were to be compared to other world zones of roughly similar ecological conditions. Thus the study involved Peru, as a specific case, and a global dimension for comparison.

Two complementary lines of investigation were:

(1) a comparative study of potato farming in the four main regions where CIP is now conducting research (station and on-farm).

(2) a global description and comparison of potato agriculture in developing countries from the perspective of agrarian ecology.

This study corresponded to a time when the Social Science Department was beginning to piece together a coherent picture from fragmentary information obtain during various studies covering potato production in different world areas. Except for CIP's statistical *Potato Atlas* (Horton, 1978), no single publication was available to give an overview of potato agriculture in developing tropical and subtropical countries. Suitable maps showing where potatoes were grown and under what kinds of farming systems were unavailable. One of CIP's roles is to gather, analyze, and distribute this kind of information in a form useful to developing countries.

The organizing scheme for the comparative study was drawn from ecological anthropology. The principal hypothesis was that similar ecological conditions will give rise to similar potato production patterns and farmer strategies. Therefore, roughly similar technologies will be applicable in similar zones. In analogous areas (tropical mountains, arid zones, lowland tropics), constraints and potentials are hypothesized to be similar, implying that each region need not be approached as totally unique. If true, this hypothesis could be very important in transfer of CIP-related technology. For example, the arid coast of Peru supports a potato farming system roughly similar to irrigated, desert potato producing regions of North Africa, and the Great Indian Desert, especially the Punjab.

Arid land potato production tends to be irrigated, commercial, dependent on imported seed, a "winter" crop, and oriented toward marketing in urban centers. Due to environment factors similar technical problems should arise from salinity in connection with irrigation water useage, for example. It also seems logical that farmer's problems with diseases, insects and pests should correlate with agro-

13

Food consumption and nutritional anthropologists have long shed new insights into biological and cultural aspects of the human diet. CIP anthropologists have brought a degree of cultural relativism to biological science thinking about consumer preferences for color, taste, shape, and cooking quality. Rwandese children (above) prepare a potato meal. (Photo Susan V. Poats.)

ecological zones. Similar patterns can be isolated in highland tropics, hilly tropics, and subtropics. However, all technology must ultimately be adapted to location specific social and economic conditions and fit the cropping system.

A "Movement of Ideas"

Anthropology has long searched for and attempted to explain regularities and parallels cutting across societies. Interest at CIP was shown in the comparative work because it illustrated to biological scientists how the Peruvian zones related to worldwide agroecological conditions. It was recognized, especially by management, that experimentation with new agricultural technology is an expensive and time-consuming process for scientists and farmers alike. Although technology has to be adapted to local ecological and cultural conditions, it is cost effective to utilize agricultural experience and knowledge gained in other areas. This is the

14

essence of the extrapolation of technological principles: movement of ideas from one area where it has been developed to one where there may be a need. Agricultural history has shown that the transfer of technology between drastically different economic or ecological systems has often resulted in limited success (temperate to tropical zones, flatlands to mountains, etc.). Thus, the comparative anthropological exercise of analyzing how potato production in similar demographic and ecological zones manifest similar technological needs is, as one biological science colleague put it, a "novel contribution to our way of thinking about agricultural technology generation and transfer."

Although it was a relatively easy task to compare in general terms the ecology of Peruvian research sites with other developing country areas, a far more difficult task centered on the comparison of potato production systems. Since potatoes are often produced in remote mountainous areas with diverse ecological characteristics, government statistics or reports rarely deal with potatoes, favoring instead commercial crops and grains.[9] As a result, the International Potato Reference files were established in which relevant production and post-harvest data were organized.

This required a global effort in data accumulation including feedback from

---

[9]The only other work similar to this on the potato in developing countries was also published by an anthropologist (see Laufer 1938).

A cross section of the central Andes showing types of region, approximate elevations, and CIP research sites.

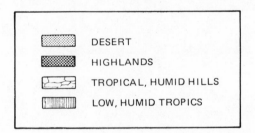

DESERT

HIGHLANDS

TROPICAL, HUMID HILLS

LOW, HUMID TROPICS

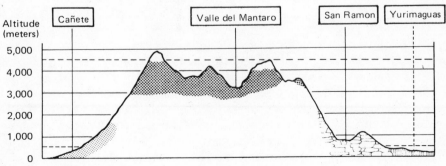

Altitude (meters)

Cañete   Valle del Mantaro   San Ramon   Yurimaguas

5,000
4,000
3,000
2,000
1,000
0

Characteristics of agricultural systems in four Peruvian regions.

Characteristics	CAÑETE	MANTARO	SAN RAMON	YURIMAGUAS
Ecological zone	Arid coast	Tropical highland	Mid-elevation, humid	Lowland, humid tropics
Farming system type	Irrigated commercial	Small-scale subsistence	Mixed shifting cultivation and tropical estate agriculture	Shifting cultivation
Principal crops or crop types	Cotton, potatoes maize	Andean tubers, grains, vegetables	Coffee, tropical fruit, cassava, maize	Rice, cassava, plantains
Method of land preparation	Plow cultivation	Plow cultivation	Clearing by fire, no tillage, plow cultivation on estates	Clearing by fire, no tillage, digging stick
Manuring or use of chemicals	Intensive	Intensive	Limited on small farms, widely used on estates	Extremely rare
Cropping pattern	Monocrop	Monocrop	Intercropping; relay planting	Intercropping; relay planting
Backyard garden	Well-defined frequently fenced	Well-defined	Dispersed tropical trees and yard plants, no fencing	Dispersed tropical fruit trees and yard plants, no fencing
Agricultural calendar: sociocultural factors	Fixed dates, government regulated	Fixed dates, community and individual decision	Dates highly variable; individual decision	Dates highly variable; individual decision
Social unit of production	Cooperatives, individual households	Individual households	Cooperatives, individual households	Individual households
Present status of potato production on farms	Modernized geared for export to urban areas	– Traditional technology – Advanced seed production	– Experiments by farmers at elevations above 1000 m. – Geared for home consumption as supplemental vegetable	Non-existent

16

CIP's regional offices, personal interviews with national program workers, and library searches. This research has taken almost 5 years but has yielded indepth information on 80 developing countries in the Americas, Africa, Asia, and the South Pacific. The files are now being analyzed, yielding information ranging from the history of the potato in each country to details of major potato diseases and pests. For the first time detailed maps of potato production in developing countries can be developed.

Over the long run, anthropology's search for parallels, principles, and cultural laws may be one of its most important contributions to technical agriculture (see also Doherty 1979; Cancian 1977). Increasingly, given limited resources, research administrators realize a shot-gun approach is not a viable option. Systematic social science information is needed so policymakers can develop relevant, timely, and workable programs for target populations. However, this will require research administrators to look at the social sciences—including anthropology—as legitimate research areas which can make useful contributions if provided adequate incentives and resources.

III. Anthropologists as Interdisciplinary Team Members: Developing Post-Harvest Technology[10]

Anthropologists have generally served in two roles assigned to them by funding agencies and agricultural development project directors. One role is to conduct a social feasibility study prior to implementation of a project and the other is to conduct an evaluation of the project after it is finished. Rarely are anthropologists incorporated into a project from beginning to end. Anthropologists, like economists, are typically thrust into the potentially unpopular role of deciding beforehand if a project is going to be worthwhile (a sensitive point to technologists promoting the project) or evaluating if it was a success or a failure (also a sensitive point if the project fails). Anthropologists assigned these roles are frequently caught between their own intellectual honesty and the strong pressures brought by what project directors and biological scientists want to hear.

At CIP an attempt has been made to develop a different approach to interdisciplinary team research. CIP's source research is organized around research "thrusts" involving collection and maintenance of a world germplasm bank, breeding to control disease and pests, agronomy, seed production and distribution, and post-harvest technology. CIP has adopted the philosophy that interdisciplinary research teams should work on these problems not in a fragmented manner but in a coordinated and continuous way. One of these teams has been composed of anthropologists and post-harvest technologists.

---

[10]This chapter is based in part on previously published materials by Rhoades et al (1982), Rhoades and Booth (1982).

CIP anthropologists conducted comparative farming systems studies in four distinct agroecological zones in Peru. This research illustrated that potatoes were produced in distinct ways in each zone and therefore required different kinds of technology. Further research explored the possibility of extrapolating research results between world zones.

At left, a CIP anthropology assistant interviews a farmer in the Amazon basin about methods, costs, and labor required to clear a jungle field.

Potatoes stored in a Himalayan house. Through their study of cultural features such as traditional use of space or farm dwelling architecture, anthroplogists bring a new dimension to post-harvest technology.

This potato store is technically excellent but was never used by farmers. Anthropologists working with CIP post-harvest technologists approached the problem of storage from the farmers' point of view. The result has been the successful adoption and adaptation of a new storage system by thousands of developing country potato farmers.

Since the late 1960s, the Peruvian government and various development agencies operating in Peru have sought technical solutions to help control the flow of consumer potatoes into the Lima market. As a result, numerous consumer potato storage facilities have been built around the country, including five large storages with a combined total capacity of 20,000 metric tons.

The largest storage complex is near the mining town of La Oroya, more than 3500 meters above sea level. These naturally-ventilated, forced-air stores were built to take advantage of the low temperatures and high humidities at high altitudes during the night (Fernandez 1976). The stores are situated roughly halfway between the major potato-producing areas of the Department of Junin and the Lima market. On initial impression, the idea behind the stores makes good sense. Potatoes could be held at La Oroya with minimum losses until prices improved in July or August in the Lima market. Theoretically, everyone gained. Farmers could get higher prices than if forced to sell immediately at harvest in May. Consumers gained as well by having to pay lower prices during the "critical months" for potatoes.

Any traveler along Peru's central highway from Lima to Huancayo, the capital of Junin Department, can visit the impressive Oroya storage complex. However, it, and the others in highland Peru built during the same period and later, today stand empty, just as they have virtually every day since they were built. These stores are existing monuments to mistargeted development projects, although according to storage specialists, they are technically sound, and, extremely well-designed. The failure came through not understanding the post-harvest system of potato agriculture as it functions in the central Andes. Such mistakes are not unique to Peru. Similar potato stores, technically sound but equally empty, can be found throughout the developing world. Against this background of 25 years of frustrating failure in attempting to improve existing post-harvest practices, anthropologists working with biological scientists have developed a new approach.

The Case of the Rustic Potato Stores

To understand the contribution of anthropology to this interdiscipinary research, it is necessary to study carefully the interaction —often conflictive— that occured between anthropologists and storage technologists. Initially, anthropologist Robert Werge set out in the Mantaro Valley to study post-harvest activities and problems facing highland potato farmers. The storage specialists at first restricted their activities to conducting research on the experiment station, also situated in the Mantaro Valley. From the beginning, however, a dialogue was maintained between team members.

Werge's ethnographic information from farmers and their practical situation soon called into question research decisions taken by post-harvest specialists on the experiment station where controlled conditions are possible.

A CIP sociologist interviews farmers on their opinions of the new diffused light store they have recently constructed.

Anthropological research at CIP has shown that farmers' perceptions of problems may be different than scientists' perceptions. In the storage work, for example, "losses" meant one thing to farmers and something else to scientists. A main concern of farmers was the loss of time and cost of labor in desprouting improved seed potatoes stored by traditional methods.

Storage Losses or No?

An intra-team debate surfaced over the concept of "storage losses," a central issue for post-harvest technologists. The potato, a vegetable tuber, is highly perishable. Post-harvest technologists were logically concerned with how to design a storage system to reduce pathological and physiological "losses," major problems in Europe and the United States. Werge, on the basis of an informal 2-month study, argued that Andean farmers did not necessarily perceive small or shriveled and spoiled potatoes as "losses" or "waste" because all potatoes were used in some form. Potatoes that could not be sold, used for seed, or immediately consumed at home were fed to animals (mainly pigs) or processed into dehydrated potatoes that could be stored for long periods. Farm women claimed that shriveled, partially spoiled potatoes tasted sweeter and were sometimes more desired (Werge, 1977).

Also an underlying assumption of earlier post-harvest technological research in the Andes was that traditional farmer storage practices were "backward," a primarily cause of "losses." Outsiders entering an Andean house have the impression of total disorder. Across the main living area hangs a string of ears of corn, against the wall next to the bed are farm tools, and below the bed are piled small, shriveled potatoes, guinea pigs scamper about the room, hiding behind the worn straw mat that holds the potatoes. It is easy to conclude, as does a recent FAO proposal calling for more storage research in the Andes, that farmer's storage practices are inadequate.

Anthropological research helped identify seed storage of improved potatoes as a farmer perceived problem in Peru and other developing countries. CIP technologists began experimentation on seed storage under naturally diffused light, a technique which aids in control of sprout growth and lessens pest and disease damage. In the photograph above, diffused light reduced sprout growth while potatoes in dark storage (left) sprouted excessively.

Virtually all technical potato storage programs developed earlier in the Andes emphasized the need for specialized structures as used in Europe or North America. Unlike in developed countries, however, potatoes in the Andes are rarely stored in separate, specialized buildings. In the early 1960s, an ethnographer (Stein 1961) noted:

> the main economic function of the house is storage of agricultural products and tools and it serves to shelter at least some of the animals as well. Its functions in sheltering people are almost secondary to the basic purposes.

The house offers security against theft and the darkened rooms hide one's wealth against the prying eyes of neighbors and employees of the agrarian bank.

Anthropologist Robert Werge (1980) later concludes:

> Concentration on specialized constructions derives from use of a model based on the contemporary European and North American practice of keeping domestic and farm activities separate in specific houses, sheds or barns. Potato farmers in developed countries have highly sophisticated storage buildings with large scale capacities, often constructed with special financing.

> This model is not appropriate to the Andes where farmers regard the storage of food, seed and tools as a domestic activity. The flexibility of space within the household residence and the security of the house are not compensated for by technical advantages which a specialized storage facility can provide.

According to Werge's study, however, farmers had "problems," but different ones than scientists had originally imagined. Farmers claimed the difficulty was not with their storage technology *per se* but with "new varieties" that produced long sprouts when stored under traditional methods. The long sprouts had to be pulled off before planting and this was considered by farmers to be labor costly. As a result of this anthropology-technical science dialogue, Werge and his colleagues concentrated on a new method of storing *improved* seed potatoes under farm conditions.[11]

Since 1972 CIP had been experimenting with a technique long known to farmers in some developing countries: natural diffused light reduces sprout elongation (Dinkel, 1963; Tupac Yupanqui, 1978). However, it was not known if the technique could be widely used in storing seed tubers under farm conditions. On the experiment station, research demonstrated clearly that indirect light reduced sprout elongation and improved overall seed quality under Andean conditions.

The design of experiment station stores, however, was from the technologists' point of view. Questions remained. Was the design relevant to farm conditions and was it acceptable to farmers? Answers could only be found through continued ethnographic research and on-farm trials with farmers acting as advisers. Werge had been doing research on the architecture and uses of farmhouses and buildings with an eye on how the diffused light principle might fit. A storage facility separate from the house did not seem realistic because of lack of security and convenience. Nor did it seem possible to introduce diffused light into dark rooms traditionally used as storage areas.

Diffused light also produces greening in potatoes, often rendering them undesirable as human food. Many small Andean farmers prefer to store all potatoes in the dark, even those to be used later for seed. This is a precaution in case of later food shortages or if they must market consumer potatoes to acquire extra income. With these socioeconomic considerations in mind, the team inspected farmhouses and talked over the problem with farmer cooperators.

Many Andean houses have a veranda or corridor with a roof that permits entrance of indirect light. The team decided to set up special seed trays as used on

---

[11]This information is based on personal communication from Dr. Werge. Interestingly, still today CIP economists and biological scientists each have their own version of the story. I expect this is inevitable in interdisciplinary research: each discipline interprets the problem in its own way and perhaps overstates or misstates the position of the other discipline. Professional "ethnocentrism" in agricultural development is still more powerful than we like to admit.

After considerable modification based on advice from farmers, the CIP interdisciplinary post-harvest team of technologists and anthropologists developed this rustic seed store model. This prototype has been promoted in 25 countries by national programs. However, as subsequent photographs illustrate, farmers rarely copied the model but adopted the idea to their own conditions and budgets.

Anthropological follow-up research on the transfer of information related to rustic seed store construction showed that virtually every farmer developed his or her own unique design based on the diffused light principle. This simple store (left) built in Peru shows some characteristics of the CIP model store but with special changes made by the farmer.

the experiment station in the houses of cooperating farmers. The trays, similar to open vegetable crates, were stacked in corridors of farm compounds with diffused light instead of direct sunlight.

These on-farm experiments gave similar scientific results as did those on the experiment station (Booth et al. 1983). Upon seeing that diffused light storage reduces sprout elongation, farmers expressed interest but were then concerned about cost of seed trays. In response, the team built simple collapsible shelves from local timber and used them in a second series of on-farm trials. The results were again positive but this time farmers were able to relate more closely to the rustic design of the stores. Throughout this adaptive process, scientists were learning more and more about technical and socioeconomic aspects of storage as well as about the proposed new technology itself.

When Werge left CIP in 1979 no evidence was available that farmers would accept the technology. The validity of the team's adaptive research approach still depended on whether farmers were willing to use the diffused light principle at their own expense. The design and initial testing of an appropriate technology is the first crucial half of the process. The second stage began after the idea of rustic, diffused light stores was introduced through CIP training courses to potato workers in Asia, Africa, and Latin America.[12]

It was during this same year (1979) that I joined CIP as its second staff anthropologist. On a trip to the Philippines, I had the opportunity to apply anthropological analysis to the first case of transfer to farmers of the idea of replacing dark seed potato stores with diffused light stores similar to those developed in Peru (Rhoades et al., 1979). A storage specialist at CIP, Robert Booth, had worked closely with national potato program workers the previous year to determine if the diffused light rustic store idea being studied in Peru was relevant to the Philippines main potato producing region in northern Luzon. As a result, the farmers in one Philippine community decided to put up a demonstration diffused-light seed store. This store was followed by five more demonstration stores built by the Philippine National Potato Program (Rhoades et al., 1979, 1982).

As Booth and I visited the area later it was clear some adoption of the technology had taken place since his first visit. To better understand farmer responses to the innovation, we developed a questionnaire to be applied by national potato program workers (Rhoades et al. 1979). In addition, in-depth interviewing of key informants provided ethnographic data detailing how the technology diffused through his or her community. This information was positively received by CIP biological scientists and helped alter the previous image of social scientists as only bearers of bad news. Subsequently, comparative studies have been conducted in several countries where the diffused light storage technology has been introduced (Rhoades et al., 1983).

Change agents from several countries involved in storage research expected that farmers would copy the demonstration stores. They had difficulty believing adoption could occur through farmers own ingenious methods of adapting an

---

[12]The special interest that Robert Werge had in training as a transfer mechanism cannot be underestimated. The post-harvest thrust has probably been the most aggressive in CIP in using training.

25

A variation of the storage idea but with potatoes over layers of *muña*, a native Andean plant that repels insects.

idea to their conditions. Anthropological follow-up in adoption areas, however, demonstrated clearly that "technology" as a unique physical "package" was not being accepted. The diffused light principle was being translated into an amazing array of farmer experimental and adopted versions of potato stores each with its own cultural flavor. After exposure to a demonstration model, farmers began to experiment on their own. For example, in Peru many farmers began by simply spreading a few potatoes under the courtyard veranda away from direct sunlight,

Many Peruvian farmers simply spread seed potatoes under a roof extending over part of an inner courtyard. This adaptation costs little and helps assure the farmer security against theft.

A farmer adaptation of the diffused light method of storage of seed potatoes. Note the traditional use of *muña*, an Andean plant that repels insects.

Another adaptation of the diffused light storage idea by an Andean farmer. (Photo courtesy Oscar Cuyabamba).

Farmers in several countries simply spread seed potatoes in front of windows to achieve the diffused light effect. This Philippine farmer built racks to improve ventilation. (Photo courtesy Mike Potts).

an experiment that involved no physical alteration of a building. Other farmers, either as a first stage adoption or elaboration of the spreading trial, constructed a simple raised platform under the veranda, a modification that allowed for better ventilation. Other farmers built simple structures, but few of these were exact copies of demonstration stores. In a few cases, associations of farmers built stores up to 100 tons capacity, many times larger than the rustic demonstration models. However, while these new physical storage structures in Africa, Asia, and Latin America reflected the unique cultural architecture of the area, the basic structural design remained similar. By 1983 more than 3,000 cases of adoption were documented.

Ethnographic information on farmer creativity in experimentation, and adaptation of the diffused light storage principle has been collected through follow-up

Known farmer adopters of diffused light seed storage.

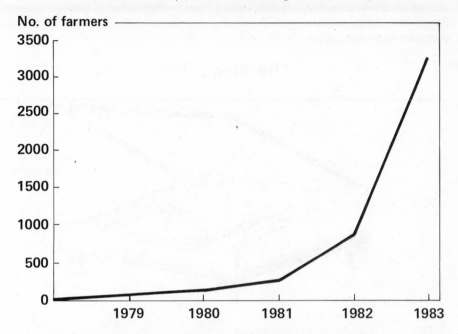

evaluations around the globe (Rhoades, et al. 1983; Rhoades 1984). It led us to urge national programs to establish demonstration stores encouraging farmer experimentation and illustrating different ways that the principle might be adapted. In many cases farmers did not automatically understand the relevancy of the principle, especially if the national program had constructed a costly demonstra-

Forty-two small farmers in the Philippines joined in building this 100-ton store. Interestingly, this store developed by farmers, is far more sophisticated than any CIP model. This illustrates how farmers adopt technology not only to technical conditions but social ones as well.

tion model. Since extension workers sometimes become frustrated when farmers did not precisely copy their design, the CIP team emphasized in training courses the anthropological concepts of cultural adaptation (Bennett 1976) and indigenous experimentation (Johnson 1972).

The Black Box

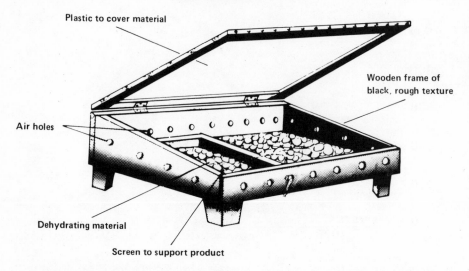

Plastic to cover material

Wooden frame of black, rough texture

Air holes

Dehydrating material

Screen to support product

Through this monitoring process, the post-harvest team learned from farmers how to improve the technology and avoid production contexts which might be inappropriate. For example, in areas where farmers want to break dormancy rapidly to meet a planting date, the diffused light principle offers few advantages. Early emphasis of what to expect and what not to expect from a new technology and by defining under what agroecological conditions a technology is likely to work or to fail, anthropology streamlines the transfer process, saving valuable time and resources needed elsewhere for agricultural development.

Potato Processing

An even more illustrative case of how anthropologists helped define and debate the directions of technology development deals with the processing component of CIP's postharvest technology research. Throughout the central Andes, potatoes are dehydrated by traditional methods of solar drying so they may be stored for up to a year or longer (Werge 1979). At the outset of the project, organizations furnishing funds and CIP processing specialists believed farmers needed a more efficient and rapid way to dehydrate potatoes. Roy Shaw, a processing specialist, in attempting to develop this more efficient method designed a simple "black box" expected to speed up the drying process. Later, Werge and Shaw took the black box directly to farm households for testing. They soon discovered that faster drying was not considered important by their informants. What farmers wanted, according to Werge, were more labor-efficient methods in cutting and peeling potatoes.

In the rugged terrain of the Andes, great distances between fields and an enormous range of farm and off-farm activities place intense pressure on family manpower (see also Brush 1977). To farmers it did not matter if 2 weeks or 2 months were required to dry potatoes. The "black box" offered no advantage over traditional methods (Werge 1979, see the appendix of this monograph for Werge's report on the black box).

Shaw, in thinking back on the experience, noted:

We were again designing postharvest technology from a distance. Since we were dealing with a dehydrated product, the problem seemed one of solar drying. We knew about peeling and cutting, but since those were labor-intensive they were thought of as desirable and not as problems.

Based on Werge's findings, Shaw reoriented his technical efforts toward developing simple peeling and cutting equipment as components of a total system of producing dehydrated potatoes. Included also were socioeconomic components: the equipment must be culturally acceptable and capable of being built by local craftsman with local materials. These components can be clearly recognized as the product of "anthropological thinking."

After Werge left CIP the post-harvest team did not entirely abandon the concept of solar drying, but looked for a context where it might fit. Only 5 of 52 families Werge studied sold part of their produce. The biological scientists assumed that demand for dehydrated potatoes among migrants from the mountains now living in coastal cities might justify a shift in scale of production of dried potatoes, the traditional *papa seca*. If it was realistic to produce dehydrated potatoes on a scale larger than the family level (village level, cooperatives, or commercial enterprises), improved solar drying efficiency would be desirable as part of a complete process. Low cost *papa seca* and starch processing plants were then designed and built with local expertise and equipment. These plants were demonstrated to possible clients in 1979-80 through field days.

In 1982, the biological scientists requested a follow-up study of the program to transfer information regarding design, construction and use of potato processing plants, presumed to have been built as a result of CIP's training activities with the Ministry of Agriculture. However, results of this evaluation were disappointing. A number of individuals had started construction of plants but due to low prices and limited demand for *papa seca* had dropped their plans. This unfortunate turn of events again placed the anthropologist in the role of "bearer of bad news," since hopes were high that the processing work would result in the same kind of success that the storage project was having. Due to heavy travel demands, the technologists on the team were unable to participate in the follow-up evaluation.

Within time, however, the major conclusions of the follow-up were accepted: demand for *papa seca* was limited and industrial or village level plants aimed at producing solely *papa seca* or potato starch were not economically viable under present price conditions. *Papa seca* is mainly consumed as an ingredient in a festival dish, *carapulcra,* only once or twice a year, almost exclusively along the coast. In the highlands, *papa seca* is produced by individual households and stored mainly for their own use. Fresh potatoes are too expensive to use for starch processing in Peru given other crops such as maize and cassava. Recommendations were made that new and creative ways of diversifying the *papa seca* plant and developing other dehydrated products that might be included in mixes or soups be sought following a serious market-demand study (Rhoades, 1982b).

The technologists felt that one potential solution to the high price of dehydrated potatoes was to develop a packaged ready-mix, nutritionally balanced, and at a reduced cost. Cheaper ingredients, such as rice and beans, helped lower cost of the potato-base mix. A potato based, ready-mix could be targeted as a weaning food for low income, nutritionally deficient groups in Lima. In light of this new direction, the social science team members (sociologist and anthropologist) conducted a feasibility study (Benavides and Rhoades, 1983). An important topic for investigation was to understand why previous attempts at formulated mixes targeted for the urban poor in Peru had failed.

Once again the post-harvest group soon found itself embroiled in an intra-team debate. The social scientists argued that according to their findings poor people of Lima's *pueblos jovenes* were already consuming the elements in the presumed packaged mix in culturally preferred forms, as dried or fresh products readily obtainable on the local market at an equal price suggested for the potato-based mix. Interest in "convenience foods" among the urban poor was yet to be empirically determined. Following a debate in which the evidence was weighed, the technologists decided from a product development point of view that the most viable option was to optimistically move ahead with the idea of a formulated mix aimed toward the urban poor. The exercise of pursuing an acceptable mix was seen as a way to learn more about consumer preferences for processed potato products and to seek ways to expand the demand for potatoes.

Despite periods of "constructive conflict," the CIP post-harvest team remains committed to the idea that technical and socioeconomic matters are equally important, even when no agreement on a project's direction can be developed (see Rhoades and Booth 1983 for a discussion of interdisciplinary team research). The approach which Booth and I jointly formulated, called by us "Farmer-Back-to-Farmer," combines anthropological and applied technological thinking (Rhoades and Booth 1982).

Farmer-Back-to-Farmer:
A Model for Generating Acceptable Agricultural Technology

The CIP postharvest team readily admits that adaptive research potentially involves at least three distinct groups each with their own perception of reality: social scientists, technologists, and farmers or other clients. Each view of reality can be considered true in and of itself and is based on the group or individual's relationship to the situation at hand. Technologists are under strong pressure by donors, administrations, and colleagues to produce a better technology that works and is adopted by farmers or consumers. Social scientists are faced with a "marginal man" or cultural broker's role: articulating their understanding of the farmers' situation to colleagues from biological sciences. Then, to complete the triangle is the farmer, the one facing the problem but who does not receive a guaranteed monthly check to "solve farmers' problems." Farmers live in both a technical and a social world based on agriculture; researchers simply study the worlds but do not have to live by the consequences of farm decisions. And all this

One of the special skills anthropologists bring to agricultural
research and development is the
ability to informally interview
farmers and place farm decisions on
technologies in social context. A
CIP anthropologist from Peru interviews an Andean farmer (right)
about her post-harvest practices.

boils down to an undeniable fact: the researcher and farmer sees the world dif-
ferently.

Briefly, the basic philosophy upon which the farmer-back-to-farmer model
rests is that successful adaptive interdisciplinary research must **BEGIN** and
END with the farmer, farm household, and community. It does not posit that
decisions as to what are important problems can be formulated on an experimental station or with a planning committee removed from the rural context and out
of touch with farm conditions. The model subsequently involves a series of
targets or goals that are logically linked by a circular and potentially recycling
pattern of four basic activities: diagnosis, identifying solutions, testing and adaptation, and farmer evaluation (Hildebrand 1978, Harwood 1979:38-40).
Research must come full circle from proper problem identification to farmer acceptance or rejection. Research, thus, is client- and problem-oriented. Research,
extension, and transfer are seen as parallel and ongoing, not sequential, dis-
jointed activities.

Understanding

The first activity in the Farmer-Back-to-Farmer model is an understanding and
learning stage. It is similar to the diagnosis stage outlined in farming systems
research, although relatively more emphasis is placed on what anthropologists
call the "emic" perspective; that is, putting oneself as much as possible into the
farmers shoes to understand how they view the problem in both technical and

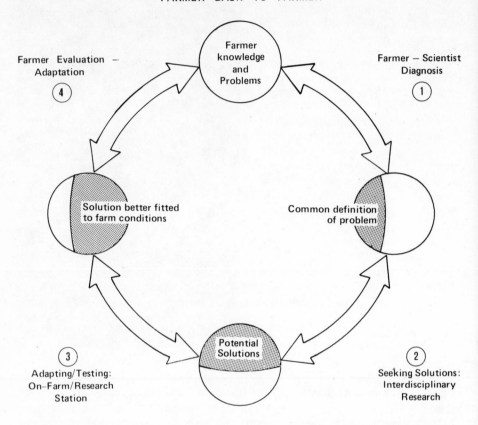

FARMER–BACK–TO–FARMER

Farmer knowledge and Problems

Farmer Evaluation – Adaptation

④

Farmer – Scientist Diagnosis

①

Solution better fitted to farm conditions

Common definition of problem

③

Adapting/Testing: On-Farm/Research Station

②

Seeking Solutions: Interdisciplinary Research

Potential Solutions

The farmer-back-to-farmer model begins and ends with the farmer. It involves four major activities each with a goal. The hatched areas in the circles indicate an increasing understanding of the technological problem area as research progress. Note that research may constantly recycle.

	Activities	Goals
1	Diagnosis	Common definition of problem by farmers and scientists
2	Interdisciplinary team research	Identify and develop a potential solution to the problem
3	On-farm testing and adaptation	Better adapt the proposed solution to farmer's conditions
4	Farmer evaluation/adaptation	Modify technology to fit local conditions; understand farmer response; monitoring adoption.

Adapted from Rhoades and Booth (1982)

34

sociocultural terms. Thus, this stage does not simply involve administering a questionnaire wherein scientists decide the relevant questions and farmers struggle to fill in the blanks. According to the "Farmer-Back-to-Farmer" approach, informal surveys or formal questionnaires are not the only early diagnostic methodological tools. Other techniques include on-farm experiments, farmer field days, farmer advisory boards, participants observation, scientists working hand in hand with farmers in their fields in exchange for information. Methods used will vary from circumstance to circumstance, depending on the special nature of transportation, time, size of region, scientists' knowledge of local conditions and populations.

The understanding stage should include farmers, social scientists, and biological scientists each using their own skills to interpret a problem area. The Farmer-Back-to-Farmer approach does not encompass specific methods for determining a ranking of constraints or priorities for agricultural policy at local or national levels but illustrates guidelines for effective design, generation, and spread of appropriate technology. Building upon, rather than replacing, traditional practices is the route to successful problem-solving.

Farmers with their long-term practical experience involving their land, mix of crops, climate and local socioeconomic conditions assume —according to the model — the status of experts in their own right and are equal members of the problem-solving team. In this beginning stage, biological scientists will naturally focus largely on technical problems. Social scientists, bound by their own selective perception, will focus on another set of phenomena: ecology, marketing, price conditions, credit restraints, or their interpretation of what farmers believe. The challenge is to weld these different perceptions into a common framework for action.

Seeking Solutions

Once the problem is generally identified and the team shares some common ground, the search for solutions is the next but perhaps more difficult stage. Despite a general assumption that a vast pool of technology is ready to be transferred to farmers, the process does not seem quite so simple. In the search for solutions, a constant on-the-spot exchange is necessary between farmers and those who test hypotheses about potential technologies of the research station. This interchange should continue throughout the selection stage. Compromises, changes, reversal of direction, or even termination of projects may be appropriate (but difficult) at this stage.

The purpose of linking on-station and farm-level team research is to arrive jointly at a definition of potential solutions, and a portion of the farmers problem always remains undefined. Proposed solutions are rarely ready at this early stage since farm problems are immensely complex, interrelated, and constantly changing.

Testing and Adapting Potential Solution

Once a solution or set of solutions is defined, the team — including extension workers if possible — should proceed to a testing and adapting activity. The objective now is to fit, with the farmer acting as adviser, the technology to local conditions. Generally, testing and adaptation occurs first on the experiment station followed by on-farm trials. Nevertheless, in the Farmer-Back-to-Farmer organization of research, even during the transfer stage, the flow of information is circular between the field and the experiment station. The technology should pass through an agronomic or technical test, an economic test and a sociocultural suitability test. The series of tests have illustrated the constant need to modify the testing methods and the technology itself. CIP's storage team, for example, began by building costly seed stores on the experiment station but data coming constantly from farmers oriented the team progressively toward less expensive designs. During this adaptive process we have found that not only does the technology change but modifications are made as well in the testing methods.

During on-farm testing, the potential solution or solutions should be compared to traditional methods. This can still be considered an understanding stage for there may be influencing factors in the farming system yet unrealized by scientists, and farmers. The testing and adaptation stage may require several recyclings to arrive at a technology worthy of demonstration and independent evaluation by farmers. And in other cases it may be possible that the traditional method cannot be improved.

On-farm research is not of much value if farmers do not feel they are a part of the research process and cannot make straight-forward suggestions on the technology under testing. This is not an easy task in those parts of the world where farmers are outwardly submissive to urban-based research scientists. Building rapport is the best way to gain farmer cooperation and this requires that scientists spend more time in the field.

Farmer Evaluation: The Crucial Stage

In agricultural development, it is a sad fact that technologies are typically released and forgotten. Storages are built, irrigation canals constructed, livestock or crop varieties introduced, all of which are rarely seen again by the innovators who, by then, have terminated their contracts and gone on to other assignments. Follow-up is rare, perhaps because the innovators assume the job is accomplished, that it is the responsibility now of a national program, or fear that the real results won't be palatable. However, according to the Farmer-Back-to-Farmer model, follow-up is the crucial final link. Data must be collected on the reception of the technology by farmers, the ultimate judges as to the appropriateness of a proposed technology.

Until this point, all scientific evaluations remain at the level of hypothesis. Unless the circle is completed, unless research results reach the farmer, prior efforts can be considered fruitless and research findings will be shelved to gather dust. And if the technology is rejected by farmers, the research process should be repeated to determine reasons and seek ways to overcome the problem. It may only be necessary to return to the adapting stage or, if the technology is totally rejected, then a new slice of the "farmer problem" needs to be taken.

The final stage involves the independent evaluation and use by farmers of the technology under their conditions, resources, and management. At this stage, scientists must not only determine acceptability but understand how farmers continue to adapt and modify the technology. Likewise, the impact of accepted technology must be monitored to make sure the technology is not detrimental to the well-being of farmers or the society at large.

Although the Farmer-Back-to-Farmer model shares many common characteristics with other farming systems models, the stress is clearly anthropological, blended with the technologists thinking. In particular, the emphasis on the farmer's point of view, informal survey techniques, and continuous adaptation and farmer experimentation fit well with anthropological approaches.

Combining on-farm experimentation with ethnographic research is an excellent way to blend anthropology and agronomy. Above a CIP agronomy assistant in Peru weighs seed for an experiment. Anthropologists subsequently conducted in-depth interviews to measure farmers' opinions of scientists' technologies.

An Andean woman selects potatoes for the market and home storage. Anthropologists at CIP have contributed to the technical program by studying often-neglected themes relevant to the generation of appropriate technology.

IV. The Anthropological Perspective: An Overview

What in fact has been the unique contribution by anthropologists to CIP's research? As social scientists, anthropologists generally cannot point to a specific technology as might a breeder point to a new variety which yields "x" tons more than traditional varieties on an experiment station. The anthropologist focuses on the clients, the people, the users of a technology, not the technology *per se*. The contribution thus comes through people-focused research and a "way of thinking" that helps identify problems and solutions followed by monitoring adaptation and acceptance by local populations of appropriate solutions. To some degree the role of anthropologists is that of "cultural brokers" between farmers and technologists. This forces the anthropologist, voluntarily or involuntarily, into a difficult "watch dog" role. If a technologically-oriented program is off base in terms of the human component, the anthropologist's role is to explain why and help steer a new course. Steering a new course, however, is not always popular, especially when programs or projects have built a momentum of their own.

Economists in technical research programs have carved out a niche based on the question: "Is a new technology or practice potentially more profitable in cash terms than the farmer's present technology or practice?" This important question is obviously more central to the analysis of technological change in fully developed market economies than in semi-commercial, subsistence systems based partially on exchange and reciprocity. With notable exceptions, the agricultural economist's model for analyzing farmer's behavior focuses on the business element of the farm enterprise, not the whole household in its ecological, physical, and sociocultural environment. Maximization of profits is thus seen as a primary human motivation.[13] The business paradigm is powerful and appealing to agricultural development specialists who themselves originate from developed, free-market economics where profit seeking is highly valued.

Economics, as broadly understood, obviously involves a great deal more than simply maximization theory (Collinson 1982). In fact, economics and anthropology have played methodological and theoretical ping-pong over the years causing sharp distinctions between the disciplines to be blurred.[14] It might be argued that the rise of Farming Systems Research reflected the growing realization that agricultural economics needed reorientation when dealing with small, subsistence farmers in developing countries (Redclift 1983). This has brought about a strong anthropological orientation of many farming systems agricultural

---

[13]As noted by Bennett and Kanel (1981), farm management grew into the field of agriculture economics which, in light of developments of the American farming context, "became increasingly focused on the farm as a business enterprise operating in the larger mileau of a capitalist national economy. Since the farm economist is concerned with the enterprise, his data are devoted to minute specifications of income, costs, yields, taxes and depreciation."

[14]This is especially true of British agricultural economics of the African farm management school. The way Collinson (1982) defines the role of the farming systems economist is very close to my definition of the anthropologists's role.

economists (Norman 1980, Hildebrand 1978). For example, Dillon and Anderson (1984) have suggested social scientists address four important matters in FSR:

1. farmers' social mileau, including customs;
2. institutional and policy setting, including land tenure, credit, and taxation;
3. economic development, including market trends and opportunity costs;
4. attitudes and personal constraints of farmers including beliefs toward change, leisure, education, different food.

The reader will note that agricultural economics in training and past research experience, especially in the United States, only has a clear comparative advantage over anthropology and rural sociology in one area, that of the economic environment. The rest, especially the first and last, clearly fall into the domain of the "other" social sciences.

Although anthropologists do not generally translate their data into dollar and cents calculations or cost-benefit ratios, they offer a human-centered perspective on farming, cost-effective methodologies, and theories that dynamically trace linkages between farm household or unit, environment, technology, crops and animals, and the larger socioeconomic milieu.[15] Anthropology thus broadens our understanding of agriculture in developing countries. CIP anthropologists, for example, stressed components that other social or biological scientists rarely consider: concepts of space, time, and farmstead architecture in the post-harvest research, dualism in farmer potato growing strategies in the ethnobotanical study, relation between agrarian ecology, cropping patterns, farm types in the Mantaro Valley project, culinary, shape, and color preferences of potato consumers in the nutrition study, and the cross-cultural comparison and technology extrapolation in the worldwide potato study.

Methodologies included participant observation (crucial in the Farmer-Back-to-Farmer model), informal surveys, techniques to understand and recognition of the importance of indigenous knowledge systems, use of aerial photos and landuse maps, controlled cross-cultural comparison, ethnographic eliciting techniques. And anthropologists brought the theoretical perspectives of functionalism, cultural ecology, substantivism, psychological and economic anthropology, and systems theory. Anthropologists at CIP have used these diverse techniques and theories to help focus on problem areas, encourage farmer involvement in design of technology, link experiment station and on-farm research, follow-up and help adapt new technology to real farm conditions.

In addition to their methods and theories for understanding and predicting human behavior, anthropologists share a general orientation which is valuable for agricultural research and is different than the orientations of sister social sciences (see Montgomery 1977:43):

1. Direct, sustained contact with the people studied in their everyday lives and on their terms, to understand how they perceive the world and their problems. This "participant observation" gives rise to the notion of "cultural relativism" which simply means "a people must be understood within the context in which

---

[15]Redclift (n.d.:31) has pointed out -- I believe correctly so -- that Farming Systems research as recently adopted by many international centers rests on principles long known by anthropologists. Anthropology therefore has much to offer this new holistic and exciting approach in terms of theory and methods.

they live," not to be judged from another context. This obviously has implications for the transfer of agricultural technology from foreign countries or from experiment stations to farms.

2. Recognition that "real" and "ideal" cultural patterns exist; that is, what people say or believe may have little relationship to what they actually do. Anthropologists argue that much of human behavior is "unconscious." While using formal questionnaires, they believe that relying only on questionnaires for information about farming practices is risky. For example, the Karimojong of Uganda, a semi-nomadic pastoral people who also farm, declare that only men and boys herd cattle while women and girls work in agriculture. Close ethnographic study, however, revealed this was the ideal behavior, not *actual* behavior. Men accounted for 35% of the labor in planting sorghum, 50% of the labor in planting millet, one-third of the labor in weeding millet fields, and over 50% of the labor during harvest (Dyson-Hudson 1972, see also Vierich 1983).

3. Recognition that a great deal of human culture and behavior is expressed in non-verbal ways (gestures, postures, use of space, modes of dress, intricate dietary patterns). For example, in the rustic storage case discussed above the anthropologist stressed the importance of local farmstead architecture in the acceptance of technology. In another case, anthropologist Mary Douglas (1974) has argued that food not only serves a biological function but also has clear esthetic functions involving color and form and is therefore similar to clothing and housing. This may explain in part the importance of color of potato varieties in consumer preferences in many developing countries. The common belief held by some agricultural scientists that poor people will eat anything (thus all that is needed is more food) may need revision in light of anthropological studies of food preferences, taboos, structured sequences of consumption, and variations in meal types. Despite prevailing common sense theories, food preferences and taste patterns are indeed among the most difficult aspects of human behavior to study.

4. Perception that all manifestations of human behavior are interrelated parts of cultural systems. This viewing or sensitivity toward relationships is the holism of anthropology. Unfortunately, "holism" has been considered a bad word in the practical world which connotes "wispy, intellectualized, convoluted explanation that fit some caricatures of the humanities" (Cancian 1977). Nevertheless, agricultural systems are in reality powerfully "holistic" and not simply made up of potatoes, soils, credit structures, etc. in isolation of each other.

Compartmentalized research in agriculture often leads to laboratory or experiment station scientists who have little knowledge about farming. A scientist may be an expert on a single diease but know little about how the disease makes or does not make a difference in the real life or practices of the farmer. While anthropology is no panacea for the distance problem between applied biological scientists and farmers, or consumers, it does offer a broader, down-to-earth analytical framework for integrating different points of view.

5. Other special orientations of anthropology as an agricultural science include:

(a) Attention to a people's past experiences in studying a problem.

In many rural areas in the developing world, decisions about what or when to plant are made not by individual farmers but through community assemblies. Anthropologists help clarify the social conditions of technology use. This field above 4,000 meters altitude is being planted by the *ticpa* system, a minimum tillage system.

(b) Recognition that in any human group, significant variations among groups and between individuals occur.

(c) Conviction that valid cross-cultural generalizations can be made, an important aspect if agriculture research findings are to be transferred between groups and nations.

Although many of the above orientations are shared with sociology, basic differences between anthropology and sociology still remain. A report of the Rockefeller Foundation (1978:5) explains:

> Sociology training programs usually have a domestic focus and tend to limit consideration of the implications of group differences within a country to a few inductive variables — class, race, and religion. They stress the use of survey data, with its greater scope for statistical proof and lesser capacity to capture unexpected events, whereas the anthropologists stress lengthy open-ended interviewing and direct observation. Special methodologies are developed for the study of group processes and of particular types of organizational structures; while anthropologists use such methods they seldom concentrate on their refinement for general application. Like the economists, the sociologists focus more on probabilistic theories for prediction of group behavior, while the anthropologists emphasize contextual explanations and predictions.

Rural sociology and other social sciences, especially human geography, have much to offer agricultural research. However, while anthropology has moved aggressively in recent years in expressing an interest in agriculture, rural sociology lost considerable ground in the 1960s and 1970s when the discipline faced a morale problem and theoretical crises from which it is only now recovering (Newby 1982).

The Anthropological Concept of Culture

Cross-cutting and underlying all of CIP's anthropological studies is the notion of "culture." A dynamic blueprint or design for living, culture is learned behavior handed down through generations so that each new cohort of babies in a society does not have to start again from scratch. To some degree, what agricultural scientists call tradition is the anthropologist's culture. Developed in the mid-19th century, culture became anthropology's key conceptual contribution to philosophical and social science thinking about man's behavior and his place in the universe (Tylor 1864). Most anthropologists agree that human culture is an adaptive, integrated, learned, and dynamic system (Radcliffe-Brown 1952). Proponents of anthropology's ecological school approach an agricultural region or technological problem with three components in mind, often placing them within the cultural systems framework.

Human Ecology[16]

At the base of all human cultures are the practices, tools, machinery, weapons, and other technologies that articulate social life with the material conditions of their habitat. Agriculture (agri-*culture)* represents a behavior-environment interaction which is distinct from hunting and gathering, shifting cultivation, and fishing. Technological practices and inventories insure survival, not only in terms of energy procurement but for protection against weather, disease, and hostile neighboring populations. Thus, when anthropologists approach a community the first question they ask is: How do these people make a living? More specifically: what is the environment like, and what technology and social patterns have they developed to exploit that environment?

---

[16]This section is based in part on a general treatment by Harris (1971). Anthropological theory is discussed in basic terms for the convenience of non-anthropologists.

Social Organization

The practical requirements of production and reproduction require development and maintenance of an orderly social life. Order in society requires proper functioning and execution of energy accumulating activities (including agriculture) needed for survival and the reproduction of the population vital for the continuation of that society. Therefore, the social life of the farming population reflects the agricultural system. This linkage is central to anthropological "holistic" thinking, and explains why technological development must also be thought of as a social process. Patterns of migration, for example, may be a powerful but silent determinant of village technology and productivity. Optimizing production the way an agro-economic team would prefer can quickly become frustrated if target villages are dependent upon extensive labor migration of the young and able-bodied. The anthropologist should be able to help put proposed technological change in context, thus casting a clearer light on potentialities.

Ideology/World View

Anthropologist Marvin Harris (1971: 146) has pointed out

while every social species (bees, ants, apes, birds, etc.) have ecological patterns and social structures, only human groups have ideologies. Ideology includes explicit and implicit knowledge, opinions, values, plans, and goals that people have about their ecological circumstance their understanding of nature, technology, production, and reproduction; their reasons for living, working, and reproducing.

Looking at farming through the eyes of farmers, a major premise of modern farming systems research, has always been a central concern of anthropology (Malinowski 1935; Redfield 1934). Enthnographers have shown that under certain conditions ideology can be as influential over agricultural change as climate or plant disease. The American ideal, for example, of the 160-acre family farm when applied to the drier American Great Plains in the late 19th century ultimately led to the Dust Bowl of the 1930s, an ecological disaster on a grand scale. In Kenya, consumers consistently reject white skinned potatoes in favor of red skinned ones. The differences in farming strategies of different ethnic groups occupying the same ecological niche testify to the role of ideology (Bennett 1969; Cole and Wolf 1974). In Sri Lanka, a belief in "evil eye" affected the design and in some cases adoption of a new storage system where the potato crop was visible (Rhoades 1984).

Agricultural scientists often forget that "agriculture" is a human-centered, controlled and manipulated process. People are at the helm. Being people they are psychological and symbolic beings, not simply organisms responding to natural conditions or profit incentives. Anthropologists through their cross-cultural perspective should be able to alert agricultural researchers of the influences of ideology, including that of scientists. Generally, ideology reflects and helps facilitate local survival, and should not necessarily be thought of as a barrier to technology improvement. Yet beliefs and attitudes can be a source of great frustration and must be considered in research.

Anthropology's potential and unique role in agriculture involves more than that of a "cultural broker" between farmers and technologists, although this role should not be slighted. Behind anthropological research at CIP, is a long intellectual history, well-founded theory, and appropriate methods, all of which have been developed through research among tribal and peasant populations now called "small or resource poor farmers," the targets of agricultural development projects.

Some economists argue that anthropologists should be brought into projects only if "unusual" or "special" problems arise (CIMMYT Economics staff 1980, Collison 1982, Simmonds 1984). A Technical Advisory Committee (TAC) of the International Agricultural Research Centers in reviewing Farming Systems Research at four centers concludes that "production economics is essential at all stages of farming systems" research while sociology and anthropology "should not be regarded as necessarily having an essential or permanent status." They may, however, have consultative roles (CGIAR 1978:64). It may well be, however, that the breadth and holism of anthropology potentially allows it to grasp more precisely than other disciplines the intricacies, interrelationships, and dynamics of local farming in developing countries. This perspective can be used in the identification and design of appropriate technologies and projects or to evaluate adoption and impact. Anthropology's problem is not the weakness of theory or methods but its voluntary or involuntary lack of contact and exposure to agricultural programs.

Peruvian women prepare chuño. Anthropological studies of traditional methods of food processing have helped guide scientific research toward appropriate solutions to technical problems.

V. The Future: Where To From Here?

Now that anthropology is beginning to have a voice, however minor, in agricultural research and development it is crucial that positive steps are taken to insure continued input of this perspective. Vernon Ruttan (1982: 42) argues that there would be a substantial payoff to increasing anthropology and sociology relative to economics at IRRI and ICRISAT.

> Anthropologists, in particular, have demonstrated a capacity to understand the dynamics of technology choice and impact at the household and village level that is highly complementary to both agronomic and economic research.

In this final section, I suggest some directions to channel anthropology toward a formal recognition of its agricultural orientation.

First, the anthropological approach to agriculture needs a name, an identity. *Agricultural Anthropology* has been put forth as a possibility (Rhoades and Rhoades, 1980). This name is important not only to identify the anthropological specialization but to precisely communicate to our colleagues that our concern is agriculture. We *are* agricultural researchers. This name is not without precedence. Agricultural anthropologists approach agriculture through their specialized perspective just as agricultural economists approach agriculture through economics, agricultural engineers through engineering, and so on. Agricultural anthropology is roughly analagous to the European disciplines of agricultural geography (as recognized in Germany) or agricultural sociology (as recognized in the Netherlands).[17]

Agricultural anthropology is the comparative, holistic, and temporal study of the human element in agricultural activity, focusing on the interactions of ecology, technology, social structure, and ideology within local and broader farming environments, and with the practical goal of responsibly applying this knowledge to improve efficiency of food production. Agricultural anthropology views agriculture neither as a mere technical process nor even as techno-economic combination, but as a complex human creation and evolutionary process that includes equally important sociocultural and ideological components in interaction with one another and the natural environment. Agricultural anthropology is broader in scope than other agricultural disciplines which focus, and rightly so, on specialized and limited problems in agriculture (Rhoades and Rhoades 1980).

Second, anthropology must continue to report on successful work, such as that at the International Potato Center or CIMMYT. More case studies are needed showing how and why anthropology has made a positive contribution to agricultural research and development projects. Without carefully documented cases, well-written, and directed toward our agricultural science colleagues, anthropology cannot claim the credibility it needs to achieve a permanent place in agricultural research. Even when anthropology functions as a critic of programs, it should be documented how anthropological input served to benefit the clients

---

[17]I am pleased to recently learn that in the Netherlands and Switzerland the term agroanthropologist is used.

of agricultural research and hence the effectiveness of the research process. Anthropology should not be pushed for anthropology's sake, but because agricultural development and its clients can benefit from the anthropological perspective. Anthropologists must demonstrate why agricultural research and development needs anthropology. To accomplish this, continued and full-time involvement is needed. The use of internships, such as the Rockefeller Foundation post-doctoral program, is an appropriate and excellent immediate solution to the problem of exposure.

Institutions wishing to employ anthropologists and other social scientists in their programs might examine closer the case of the International Potato Center. Mentioned earlier in this report are some circumstantial and personality aspects that helped secure a role for anthropology at CIP. Anthropology, however, succeeded not simply because of "good people." The quality of anthropologists was a necessary but not sufficient cause of success. Other international centers had equally capable, if not superior, anthropologists.

Additional factors underlying success are related to the CIP organization of research and leadership:

1. CIP is characterized by a built-in flexibility which allowed anthropologists to freely seek topics where their expertise might be utilized. Anthropologists were not brought in and mandated or assigned only the role, for example, of helping economists or agronomists to understand social factors surrounding their on-farm trials. Anthropologists at CIP can more or less define their own projects as long as they are relevant to the applied job at hand. In this regard, anthropology enjoys the kind of freedom that early economists in the CGIAR system experienced, an important factor in the later successes of economics (Ruttan 1982:309).

2. CIP is a problem-oriented international center more interested in applied results than with science *per se*. If it works, use it. Thus, anthropologists were given a chance as equal partners with economists and biological scientists in interdisciplinary team research. Also team research was clearly focused on a specific crop and subsystem within potato agriculture. There was never the difficulty of the anthropologist wanting to "study everything in a few villages." Instead, research immediately centered on potato post-harvest technology (processing and storage), potato consumption, or varieties. Within these areas the anthropologist could, if desired, study "everything" as long as it related to potatoes, development, and improving technology.

3. CIP leadership also was aware of the richness of the social science disciplines other than economics. For this reason, CIP has the only Social Science Department (as opposed to economics) in the CGIAR system. Rural sociologist Dr. Gelia Castillo served on the CIP Board of Trustees since 1978 and when her board term ended she was its chairwoman.

These structural characteristics of organization combined with the personality and circumstancial events explain why anthropology at CIP has succeeded. The CIP case might provide insights for other agricultural organizations interested in incorporating the anthropological perspective into the design and generation of technology. If anthropology is viewed as a weaker sister and service role to strictly economics programs, the chances of success are probably greatly reduced. Ruttan (1982:309) in reflecting on the experiences of social scientists in agricultural research notes: "Where a social science staff has been cast in a purely service-oriented role, however, low staff morale and difficulty in retaining an effective social science capacity have tended to result."

Third, anthropology needs to shake stereotypes that follow the discipline. One major image problem, and not entirely unjustified, is that anthropologists concentrate on "traditional" aspects of farming to the neglect of the market or "modern" sector. The bias toward the noble savage living in permanent harmony with nature still affects the thinking of some anthropologists.

Understandably, any new discipline in a multidisciplinary research institute will have an image problem. This is especially true of anthropology which means different things to different nationalities, if it means anything at all. Thus a CIP Englishman asks "what is a scientist who plays around with ancient bones in basement laboratories doing in agricultural research?" An American colleague asks "what do monkey specialists have to do with an agricultural center?" Still other nationalities identify us with the study of remaining "exotic savages," who do not practice agriculture. Our thing should be headhunting, not agriculture. Even fellow social scientists hold stereotypic views of anthropologists. Unfortunately, all stereotypes carry some element of truth. Anthropologists have always had a flare for the popular so it is little wonder that outsiders think of us as students of the exotic, quaint, and outdated. The practical relevance is not always clear.

While anthropologists obviously need to sharpen their focus on agricultural problems, some criticisms against anthropology are less understandable. Anthropologists obviously commit sins, such as writing wordy reports and spending too much time in fieldwork. However, other agricultural scientists commit the same sins, albeit in different forms and less recognized. Any potato breeder will tell you it takes a minimum of 10 years to breed a new variety. CIP's storage experiments have been going on for a decade (8 years of on-farm research in the same valley). Economists at the International Crops Research Institute for the Semi-Arid Tropics in India have been researching the same few villages for over a decade. Agricultural research under any disciplinary label is a time consuming process. Anthropologists also do not have a copyright on lengthy reports. I learned this culling through over 400 technical documents on potato production, many of which contain thick reams of pages on agronomic or biological experiments that few people will ever read. It may be, however, that anthropologists and sociologists — in their minority roles — are more vulnerable to criticisms when they commit the common sins of agricultural science.

Fourth, anthropological methods are perceived as difficult to replicate and have not been clearly explained to our fellow scientists. It is often charged that the scientific method is not clearly followed, no set of working hypotheses can be noted. Research is site specific and descriptive. Since random sampling is sometimes not used, critics charge that valid generalizations cannot be drawn. Anthropologist's earlier tendency to play down quantification, statistical methods, and clearly articulated models continues to affect anthropology's credibility among agricultural scientists.

Anthropologists sometimes tend to view society as static and concentrate their research on small groups (villages, tribes) and ignore communities that are fully integrated into monetarized economics. Also anthropologists in the past have largely been concerned with reporting what exists or has existed, and have not developed models useful for predicting agricultural change. This, however, is not true of cultural ecological and economic anthropological studies of agriculture (Raintree 1984). Methods used by anthropologists, nevertheless, need to be better explained and the obvious erroneous assumption that anthropologists do not use quantification should be corrected.[18]

Fifth, anthropologists need to create a formal framework, or at least incipient structures at several levels that will push the discipline forward. Anthropology needs to formally recognize its agricultural orientation. I suggest we need to operate on two fronts: (1) promotion of the field and continued demonstration of why agriculture needs anthropology; (2) training anthropologists to qualify and compete as agricultural scientists.

An association or organization is needed that will function to promote the field and serve as organizational pivot for interested persons. One possible organization to spin this development is the U.S.-based Anthropological Study Group of Agrarian Systems (ASGAS). Its already excellent work might be further intensified, however, if ASGAS could broaden its membership to include more international members, especially from developing countries. In university training, agricultural anthropology needs its analog to the now flourishing programs in medical and nutritional anthropology. Although over the long haul we may see some formal recognition of agricultural anthropology, this is unlikely to happen in the near future. The most we can realistically hope for now are "bridger" programs between anthropology departments and agricultural schools. However, in the absence of established programs much can still be done to link agriculture and anthropology through degrees and course work combinations. An agricultural degree (preferably a degree in a technical area) combined with a doctorate in anthropology would be highly applicable. Such dual professional status is extremely attractive to potential employers, especially those in agricultural development (see Rockefeller Foundation 1978:18-42 for more information on training of anthropologists for work in agriculture).

One positive fact is that agricultural organizations do not harbor the same elitist attitude toward degrees as do academic anthropology departments. Rather than being "name-school conscious" agricultural organizations tend to be more interested in an agricultural background (in the U.S. most likely obtained in a land-grant college or through practical experience) than a "prestigious" anthropology degree. In the United States, the best possibilities for studying agricultural anthropology are at the University of Arizona, University of Florida, and University of Kentucky. Regardless of university, the student will not necessarily benefit from a narrowly focused, traditional anthropology program. Although anthropology should not give up holism, shifts in study concentration may be appropriate. Instead of learning structural linguistics or the fossil record

---

[18]This does not mean anthropologists should become enamored with empiricism for empiricism's sake. Agricultural reports already have enough statistics and numbers jerked out of their socioeconomic and time contexts to last for ages. A balanced approach is needed.

in detail, a student of agricultural anthropology could better meet his or her needs by comprehending the principles of agronomy or plant genetics. Clearly, agricultural anthropologists need to be capable of handling technical parts of agriculture.

At this early stage of anthropology's bid to become one of the agricultural sciences, no guarantees can be written for future employment. Presently, few anthropologists are employed or even specifically trained to work in agriculture-related jobs. Perhaps as many as a quarter million people work in agriculture for the USDA, USAID, international and national research centers, and FAO. The number of full-time anthropologists employed in agricultural jobs in all of these organizations could probably be counted on two hands.[19] For example, out of 736 senior scientists employed in 1983 in CGIAR, *three* are anthropologists. And CGIAR prides itself on the use of anthropologists and sociologists! This is, in my opinion, a professional tragedy. Still, given the present dismal employment opportunities with a "pure" anthropology degree, the backing of an agricultural degree will be useful on the job market.[20] What is certain is that many agricultural organizations are more receptive toward anthropologists than at any time since the 1930s.

Anthropologists who bemoan the increasing fragmentation of anthropology may find the proposal for another subdisciplinary specialization disconcerting. The university and "academically" oriented may feel that a call for the application of anthropology to planned agricultural change may violate the intellectual detachment needed for objective research. However, agricultural anthropology can be understood as both a research and an applied field, broad enough to accomodate all anthropologists who work on agricultural-related problems throughout the world. Pure research must be as central to agricultural anthropology as applied research. The methods should be flexible to match the complexity of the subject matter. And anthropologists need to fill gaps not covered by other disciplines, including economics. Anthropology's ability to deal in a solid way with the important technical, ecological, and socioeconomic aspects neglected by the other agricultural sciences will be our most valuable asset.

In conclusion, anthropology as a discipline has more than a century of direct experience in agriculture, and an intimate association with farmers in every corner of the globe. Work at the International Potato Center is an applied outgrowth of this history and empirically illustrates that anthropology can play an important role in agricultural research and development. It is now up to anthropologists to formally recognize their agricultural roots and aggressively but professionally become involved in agricultural research. Anthropologists only lack formal recognition of their long experience with farming peoples and the will to articulate their expertise to the non-anthropological world, especially to other agricultural scientists.

[19]Van Dusseldorp (1977) estimates that out of every 1,000 permanent scientists in agricultural research centers only one is a sociologist or anthropologist.

[20]At the University of Arizona, for example, two 1984 Ph.D. Anthropology graduates prepared themselves to work in food-related areas. One student received three job offers from international organizations while the other received a university-based research and teaching position. Cultural anthropology students who had no special training in agriculture did not fare so well.

REFERENCES

Almy, Susan W. 1977. Anthropologists and Development Agencies. *American Anthropologist* 79:280-292.

Almy, Susan W. 1984. Personal communication.

Barlett, Peggy F. 1980. Adaptive Strategies in Peasant Agricultural Production. *Annual Reviews of Anthropology*. Palo Alto, California: Annual Reviews Inc.

Beals, Alan. 1973. *Culture in Process*. New York: Holt, Rinehart, and Wiston.

Benavides, M. and R. Rhoades. 1982. Socioeconomic Conditions, Food Habits, and Formulated Food Programs in the *Pueblos Jovenes* of Lima, Peru. A Preliminary Study. (Unpublished).

Bennett, J.W. 1969. *Northern Plains: Adaptive Strategy an Agrarian Life*. Chicago: Aldine.

Bennett, J.W. 1976. Anticipation, Adaptation, and the Concept of Culture in Anthropology. *Science* 192:847-852.

Bennett J. and D. Kanel. 1981. Agricultural Economics and Economic Anthropology: Confrontation and Accommodation. Preliminary undocumented draft prepared for oral presentation at Inaugural Conference of Society for Anthropological Economics, at Indiana University. April, 1981.

Bidegaray, P. 1981. Agricultura en la Selva Peruana: El Caso de Yurimaguas. Lima, Peru: International Potato Center (manuscript).

Booth, R., R. Shaw, and A. Tupac-Yupanqui. 1983. Use of Natural Diffused Light for the Storage of Seed Tubers. In W.J. Hooker, ed. *Research for the Potato in the Year 2000*. Lima, Peru: International Potato Center. pp. 65-66.

Brush, S.B. 1977. The Myth of the Idle Peasant: Employment in a Subsistence Economy. In Halperin, R., Dow, J., eds. *Peasant Livelihood: Studies in Economic Anthropology and Cultural Ecology*. New York: St. Martins.

Brush, S., H. Carney, and Z. Huaman. 1981. Dynamics of Andean Potato Agriculture, *Economic Botany* 35(1):70-88.

Cancian, Frank. 1977. Can anthropology help agricultural development? *Culture and Agriculture* No. 2 pp. 1-8.

CIMMYT Economics Staff. 1980. Assessing Farmer Needs in Designing Agricultural Technology. IADS Occasional paper.

Cole, John and Eric Wolf. 1974. *The Hidden Frontier: Ecology and Ethnicity in an Alpine Valley*. New York: Academic Press.

Collinson, M.P. 1982. Farming Systems Research in Eastern Africa: The Experience of CIMMYT and Some National Agricultural Research Services, 1976-81. MSU International Development Paper No. 3. East Lansing, Michigan: Department of Agricultural Economics. Michigan State University.

Conklin, H.C. 1954. An Ethnobotanical Approach to Shifting Agriculture. *Transactions of the New York Academy of Sciences*, ser. 2, 17: 133-42.

Conklin, H.C. 1957. Hanunoo Agriculture in the Philippines. Forestry Development Paper No. 12. Rome: FAO.

Consultative Group on International Agricultural Research (Technical Advisory Committee) 1976. Report of the First TAC Quinquennial Review of the International Potato Center (CIP). TAC Secretariat. Rome: FAO.

Consultative Group on International Agricultural Research (Technical Advisory Committee). 1978. Farming Systems Research at the International Agricultural Research Centers. Rome: TAC Secretariat, Agriculture Department, FAO.

Consultative Group on International Agricultural Research. 1980. CGIAR Secretariat. Washington, D.C.: CGIAR Secretariat.

De Schilippe, Pierre. 1956. *Shifting Cultivation in Africa.* London: Routledge and Kegan Paul.

Dinkel, D. 1963. Light-Induced Inhibition of Potato Tuber Sprouting. *Science*, 141: 1047-8.

Dillon, J.L. and J.R. Anderson. 1984. Concept and Practice of Farming Systems Research. In J.V. Martin, ed., Proceedings of ACIAR Consultation on Agricultural Research in Eastern Africa. Camberra: ACIAR. In press.

Doherty, V.S. 1979. Human Nature and the Design of Agricultural Technology. Paper presented at Workshop on Socioeconomic Contraints to Development of Semi-Arid Tropical Agriculture. Hyderabad, India: ICRISAT. 19-23 February.

Douglas, Mary. 1974. Food as Art Form. London: Studio International. 188,969:83-88.

Dyson-Hudson, R. 1972. Pastoralism: Self-image and Behavioral Reality. In Irons, W. and Dyson-Hudson, N. ed. Perspectives on Nomadism. Leiden, Netherlands. J. Brill, 30-47.

Eastman, C. 1977. Technological Change and Food Production: General Perspectives and the Specific Case of Potatoes. Social Science Department Special Publications. Lima, Peru: International Potato Center.

Fernandez, Angel. 1976. Infraestructura para la Comercializacion del Producto Alimenticio Papa: Informe Preliminar. Lima, Peru: Editora Peruana.

Foster, G. 1969. Applied Anthropology. Boston: Little Brown and Company.

Flannery, K.V. 1965. The Ecology of Food Production in Mesopotamia. Science 147: 1247-56.

Geertz, Clifford. 1963. Agricultural Involution: The Process of Ecological Change in Indonesia. Berkeley: University of California Press for the Association of Asian Studies.

Goldschmidt, Walter. 1947. As you Sow. New York: Harcourt Brace.

Goldschmidt, Walter. 1978. As you Sow. Three Studies in the Social Consequences of Agribusiness. Montclair, New Jersey: Allanheld, Osmum.

Hall, Edward. 1961. The Silent Language. New York: Fawcett Publications.

Hall, Edward. 1966. The Hidden Dimension. Garden City, N.Y.: Doubleday.

Harris, Marvin. 1971. Culture, Man, and Nature. New York: Thomas Y. Crowell Company.

Harwood, Richard R. 1979. Small Farm Development. Understanding and Improving Farming Systems in the Humid Tropics. Boulder, Col.: Westview Press.

Hawkes, J.G. 1947. On the Origin and Meaning of South American Indian Potato Names. Journal of the Linnean Society (Botany). 53:205-250.

Hildebrand, P.E. 1978. Generating Technology for Traditional Farmers: A Multidisciplinary Methodology. Asian Report No. 8, New Delhi, India: CIMMYT.

Holmberg, A., M.C. Vasquez, P.L. Doughty, J.O. Alers, H.F. Dobyns, and H.D. Lasswell. 1965. The Vicos Case: Peasant Society in Transition. The American Behavioral Scientist 8(7): 3-33 (special issue).

Holmberg, A., H. Dobyns, C. Monge, M.C. Vasquez, and Harold D. Lasswell. 1962. Community and Regional Development: The Joint Cornell - Peru Experiment. Human Organization 21: 107-124.

Horton D. 1978. Potato Atlas. Lima, Peru: International Potato Center.

Horton D. 1984. Social Scientists in Agricultural Research. The Mantaro Valley Project. Ottawa: International Development Research Centre.

Janik, J., R. Schery, F. Woods and V. Ruttan, Eds. 1970. Plant Agriculture. Readings from Scientific American. San Francisco: W.H. Freeman and Company.

Johnson, Allan W. 1972. Individuality and Experimentation in Traditional Agriculture. Human Ecology. 1(2):43-47.

La Barre, W. 1947. Potato Taxonomy among the Aymara Indians of Bolivia. Acta Americana 5(1-2):83-103.

Loomis, Charles. 1943. Applied Anthropology in Latin America. Applied Anthropology 2(2):33-35.

Mayer, Enrique. 1979. Land-Use in the Andes: Ecology and Agriculture in the Mantaro Valley of Peru with Special Reference to Potatoes. Social Science Department Special Publication. Lima, Peru, International Potato Center.

Montgomery, Edward and John W. Bennett. 1979. Anthropological Studies of Food and and Nutrition. The 1940s and the 1970s. In The Uses of Anthropology. W. Goldschmidt, ed. Washington, D.C. American Anthropological Association.

Netting, Robert. 1974. Agrarian Ecology. Palo Alto, California: *Annual Review of Anthropology.* Vol. 3:21-56.

Newby, Howard. 1982. Rural Sociology and Its Relevance to the Agricultural Economist: A Review. *Journal of Agricultural Economics* 33(2):125-165.

Norman, D.W. 1980. Farming System Approach: Relevancy for the Small Farmer. Rural Development Paper 5. East Lansing Michigan State University: Dept. of Agricultural Economics.

Poats, S. 1983. Beyond the Farmer: Potato Consumption in the Tropics. In W.J. Hooker, ed. *"Research for the Potato in the Year 2000."* Proceedings of the International Congress in Celebration of the Tenth Anniversary of the International Potato Center. Lima, Peru: International Potato Center. Lima. pp. 10-17.

Radcliffe-Brown, A.R. 1952. *Structure and Function in Primitive Society; Essays and Addresses.* London: Cohen and West.

Raintree, J.B. 1984. Bioeconomic Considerations in the Design of Agroforestry Cropping Systems. *Plant Research and Agroforestry.* ICRAF Reprint 11. Nairobi, Kenya: International Centre for Research on Agroforestry.

Recharte, Jorge. 1981. Los Limites Socioecologicos del Crecimiento Agricola en la Ceja de Selva. Pontificia Universidad Catolica del Peru. Thesis. pp. 208.

Redclift, Michael. 1983. Production Programs for Small Farmers: Plan Puebla as Myth and Reality. *Economic Development and Cultural Change* 31(3):551-570.

Redfield, R. and W. Lloyd Warner. 1940. Cultural Anthropology and Modern Agriculture. In Farmers in a Changing World. *Yearbook of Agriculture 1940.* Washington, D.C.: US Government Printing Office. pp. 983-993.

Reed, Charles A. ed. 1977. *Origins of Agriculture.* World Anthropology Series. Chicago: Aldine.

Rhoades, 1982a. The Art of the Informal Agricultural Survey. Social Science Training Document Series 1982-2. Lima, Peru: International Potato Center.

Rhoades, R. 1982b. Follow-up of Transfer of CIP Processing Technology. Manuscript International Potato Center.

Rhoades, R. 1984. Changing a Post-harvest System: The Case of Diffused Light Stores in Sri Lanka. SSD Working Paper 1984-1. Lima, Peru: International Potato Center.

Rhoades R. and R. Booth. 1982. Farmer-Back-to-Farmer: A Model for Generating Acceptable Agricultural Technology. *Agricultural Administration.* 11:127-137.

Rhoades, R. and R. Booth. 1983. Interdisciplinary Teams in Agriculture Research and Development. *Culture and Agriculture.* Issue 20. Summer. pp. 1-7.

Rhoades, R. and V. Rhoades. 1980. Agricultural Anthropology: A Call for the Establishment of a New Professional Speciality. *Practicing Anthropology* 2(4): 10-12/28.

Rhoades, R., R. Booth, and M. Potts. 1983. Farmer Acceptance of Improved Potato Storage Practices in Developing Countries. *Outlook on Agriculture* 12(1):12-16.

Rhoades, R., R. Booth, R. Shaw, and R. Werge. 1982. Interdisciplinary Development and Transfer of Post-harvest Technology at the International Potato Center. In International Rice Research Institute report of an Exploratory Workshop on: The Role of Anthropologists and Other Social Scientists in Interdisciplinary Teams Developing Improved Food Production Technology. Los Banos, Laguna, Philippines. pp. 1-8.

Rhoades, R.E., R. Booth, F. Rutab, O. Sano, and L.J. Harmsworth. 1979. The Acceptance of Improved Potato Storage Practices by Philippine Farmers: A Preliminary Study. Mimeograph. The International Potato Center, Lima, Peru: International Potato Center.

Rockefeller Foundation. 1978. Society, Culture, and Agriculture. Working Paper. Workshop on Training Programs Combining Anthropology and Sociology with the Agricultural Sciences. New York: Rockefeller Foundation.

Ruttan, Vernon. 1982. Agricultural Research Policy. Minneapolis, Mn.: University of Minnesota Press.

Ryan, James G. 1979. Comment: Technology for Semiarid Northeast Brazil. In *Economics and the Design of Small Farmer Technology.* Ames, Iowa: Iowa State University Press, (First Edition). pp. 119-121.

Stein, W. 1961. *Hualcani: Life in the Highlands of Peru.* Ithaca, New York; Cornell University Press.

Thurow, Lester C. 1977. Economics, 1977. *Daedalus* (Fall) pp.79-94.

Tupac Yupanqui, A.L.1978. Aspectos Fisiologicos del Almacenamiento de Tuberculos - Semilla de Papa: Influencia de la Temperatura y la Luz. Tesis. Lima, Peru: Universidad Nacional Agraria, La Molina.

Tylor, Edward B. 1871. *Primitive Culture: Researches into the Development of Mythology, Philosophy, Religion, Language, Art and Custom.* London: J. Murray.

Van Dusseldorp, D.B.W.M. 1977. Some thoughts on the role of social sciences in the agricultural research centers in developing countries. *Netherland Journal of Agricultural Sciences* 25:213-228.

Vierich, Helga. 1984. Accomodation or participation? Communication Problems. In Coming Full Circle. Peter Matlon, et al., eds. Ottawa, Canada: International Development Research Centre.

Werge, Robert. 1977. Anthropology and Agricultural Research: The Case of Potato Anthropology. Lima, Peru: International Potato Center.

Werge, Robert. 1977. Potato Storage Systems in the Mantaro Valley Region of Peru. Social Science Department. Lima, Peru: International Potato Center.

Werge, Robert. 1978. Social Science Training for Regional Agricultural Development. Paper presented at the Meetings of the Society for Applied Anthropology. Merida, Mexico. 2-9 April, 1978.

Werge, Robert. 1979. Potato Processing in the Central Highlands of Peru. *Ecology of Food and Nutrition.* 7: 229-234.

Werge, Robert. 1980. Potatoes, Peasants and Development Projects: A Sociocultural Perspective from the Andes. (Unpublished).

Whyte, William Foote. 1977. Seed Production Systems: Notes for a Colombia Project. (Unpublished).

Whyte, William Foote. 1984. Personal Communication.

Wolf, Eric. 1964. Anthropology. Englewood Cliffs: Prentice-Hall.

ANNEX

ANNEX 1
PUBLICATIONS BY ANTHROPOLOGISTS AND SOCIOLOGISTS AT CIP

I. Working Papers

1979-2 WERGE, R., and M. BENAVIDES. 1979. Investigation of Farming in Peru by Means of a Multiple Visit Survey. Lima, International Potato Center. 14p. Working Paper 1979-2.

1979-4 WERGE, R. 1979. The Agricultural Strategy of Rural Households in Three Ecological Zones of the Central Andes. Lima, International Potato Center. 27p. Working Paper 1979-4. Reprint 1981.

1980-1 HORTON, D., F. TARDIEU, M. BENAVIDES, L. TOMASSINI, and P. ACCATINO. 1980. Tecnologia de la Producción de Papa en el Valle del Mantaro, Perú. Resultados de una Encuesta Agro-Económica de Visita Multiple. Lima, Centro Internacional de la Papa. 68p. Documento de Trabajo 1980-1.

1980-3 CARNEY, H.J. 1980. Diversity, Distribution and Peasant Selection of Indigenous Potato Varieties in the Mantaro Valley, Peru: A Biocultural Evolutionary Process. Lima, International Potato Center. 19p. Working Paper 1980-3.

1980-5 BRUSH, S., H.J. CARNEY, and Z. HUAMAN. 1980. The Dynamics of Andean Potato Agriculture. International Potato Center, Lima, Peru. 27p. Working Paper 1980-5.

1982-1 RHOADES R., and R. BOOTH. 1982. Farmer-Back-to-Farmer: A Model for Generating Acceptable Agricultural Technology. Lima, International Potato Center. 15p. Working Paper 1982-1.

1982-3 POATS S. 1982. Potato Preferences: A Preliminary Examination. Lima, International Potato Center. 18p. Working Paper 1982-3.

II. Training Documents

1980-1 WERGE, R. 1980. Social Science Training for Agricultural Development: The Case of CIP. Lima, International Potato Center. 8p. Training Document 1980-1.

1980-5 WERGE, R. and M. BENAVIDES. 1980. Investigtion of Farming in Peru by Means of a Multiple Visit Survey. Lima, International Potato Center. 12p. Training Document 1980-5. Reprint 1981.

1982-2 RHOADES, R. 1982. The Art of the Informal Agricultural Survey. Lima, International Potato Center. 40p. Training Document 1982-2.

1982-3 RHOADES, R. 1982. Understanding Small Farmers: Sociocultural Perspectives on Experimental Farm Trials. Lima, International Potato Center. 9p. Training Document 1982-3.

1982-7 RHOADES, R. 1983. El Arte de la Encuesta Informal Agricola. Lima, Centro Internacional de la Papa. 38p. Documento de Entrenamiento 1982-7.

1982-8 RHOADES, R. 1983. Para Comprender a los Pequeños Agricultores: Perspectivas Socioculturales de la Investigación Agricola. Lima, Centro Internacional de la Papa. 9p. Documento de Entrenamiento 1982-8.

III. Special Publications and Monographs

EASTMAN, C. 1977. Technological Change and Food Production: General Perspectives and the Specific Case of Potatoes. Lima, International Potato Center. 27p.

MAYER, E. 1980. Land Use in the Andes. Ecology and Agriculture in the Mantaro Valley of Peru with Special Reference to Potatoes. Lima, International Potato Center. 115p.

MAYER, E. 1981. Uso de la Tierra en los Andes. Ecologia y Agricultura en el Valle del Mantaro del Perú con Referencia Especial a la Papa. Lima, Centro Internacional de la Papa. 127p.

WERGE, R. 1977. Potato Storage Systems in the Mantaro Valley Region of Peru. Lima, International Potato Center. 49p.

WERGE, R.   1980. Sistemas de Almacenamiento de Papa en la Región del Valle del Mantaro, Perú. Lima, Centro Internacional de la Papa. 45p.

WERGE, R.   1977. Socioeconmic Aspects of the Production and Utilization of Potatoes in Peru: A Bibliography. Lima, International Potato Center. 73p. (In English and Spanish).

IV. Theses

BENAVIDES, M.   1981. Aspectos Socio-Economicos de la Producción de Papa en la Unidad Campesina (Valle del Mantaro). Tesis de Bs. Sociologia. Pontificia Universidad Católica del Perú. 70 p.

RECHARTE, J.   1981. Los Limites Socioecológicos del Crecimiento Agricola en la Ceja de Selva. Tesis Lic. Antropologia. Pontificia Universidad Católica del Perú. 208p.

V. Articles and Conference Papers

BRUSH, S., H.J. CARNEY, and Z. HUAMAN. 1981. "Dynamics of Andean Potato Agriculture." *Economic Botany.* 35(1): 70-88.

POATS, S. 1983. Beyond the Farmer: Potato Consumption in the Tropics. In W.J. Hooker, ed. *"Research for the Potato in the Year 2000,"* Proceedings of the International Congress in Celebration of the Tenth Anniversary of the International Potato Center, held at Lima, Peru, 22-27 February, 1982. Lima, International Potato Center. pp. 10-17.

POATS, S. "Más Alla del Agricultor: El Consumo de la Papa en el Mundo Tropical." Forthcoming in *Archivos Latinoamericanos de Nutrición.*

RECHARTE, J.   1982. Prosperidad y Pobreza en la Agricultura de la Ceja de Selva. El Valle de Chanchamayo. In Centro de Investigación y Promoción Amazónica. *Colonizacion en la Amazonia.* Lima, Perú. pp.105-161.

RHOADES, R. Farming on High: Andean Ecology and Aldo Leopold. Forthcoming in Michael Tobias, ed. *Mountain People.* University of Oklahoma Press.

RHOADES, R.   1982. "The Incredible Potato." *National Geographic.* 161(5): 668-694.

RHOADES, R. 1982. Toward an Understanding of Hot, Humid Tropical Farming Systems with Emphasis on the Potato. In L.J. Harmsworth, J.A.T. Woodford, and M.E. Marvel. *Potato Production in the Humid Tropics.* Proceedings of the Third International Symposium on Potato Production for the Southeast Asian and Pacific Regions, held 12-17 October 1980 at Bandung, Indonesia. Los Baños, Laguna, Philippines, International Potato Center Region VII. pp. 444-455.

RHOADES, R. and R. BOOTH.   1983. Acceptance of Diffused Light Potato Seed Storage Technology in Developing Countries. In W.J. Hooker, ed. *"Research for the Potato in the Year 2000,"* Proceedings of the International Congress in Celebration of the Tenth Anniversary of the International Potato Center, held at Lima, Peru, 22-27 February, 1982. Lima, International Potato Center. pp. 160-161.

RHOADES, R., and R. BOOTH.   1982. "Farmer-Back-to-Farmer: A Model for Generating Acceptable Agricultural Technology." *Agricultural Administration* 11: 127-137.

RHOADES, R, R. BOOTH, and M. POTTS.   1983. "Farmer Acceptance of Improved Potato Storage Practices in Developing Countries." *Outlook on Agriculture* 12(1): 12-16.

RHOADES, R., R. BOOTH, R. SHAW, and R. WERGE.   1982. Interdisciplinary Development and Transfer of Postharvest Technology at the International Potato Center. In International Rice Research Institute. Report of an Exploratory Workshop on: The Role of Anthropologists and Other Social Scientists in Interdisciplinary Teams Developing Improved Food Production Technology. Los Baños, Laguna, Philippines. pp. 1-8.

RHOADES, R. & V.N. RHOADES.   1980. "Agricultural Anthropology: A Call for the Establishment of a New Professional Specialty." *Practicing Anthropology* 2(4): 10-12/28.

RHOADES, R. Understanding Small Farmers in Developing Countries: Sociocultural Perspectives on Agronomic Farm Trials. Forthcoming: *Journal of Agronomic Education.*

RHOADES, R. Informal Survey Methods for Farming Systems Research. Forthcoming: *Human Organization.*

RHOADES, R. and R. BOOTH. Social and Biological Science Interdisciplinary Team Reserach in Agricultural Research and Development. *Culture and Agriculture.* Issue 20, Summer. pp. 1-7.

SHAW, R., R. BOOTH, and R. RHOADES. 1982. "On the Development of Appropriate Technology: A Case of Post-Harvest Potatoes." *Food Technology,* October, pp. 114 and 116-118.

WERGE, R. 1979. "Potato Processing in Central Highlands of Peru." *Ecology of Food and Nutrition* 7: 229-234.

ANNEX 2

Anthropologists and Sociologists Associated with the International Potato Center 1975-1984

Dates	Name	Discipline	Degree	Nationality	Research
1975-79	Robert Werge	Anthropologist	Ph.D.	USA	Post-harvest technology
1977	William F. Whyte	Anthropologist	Ph.D.	USA	Colombian Potato Seed Systems
1978	Enrique Mayer	Anthropologist	Ph.D.	Peru	Land-use
1977-78	Stephen Brush	Anthropologist	Ph.D.	USA	Potato Folk Taxonomy
1977-78	Heath Carney	Anthropologist	B.S.	USA	Potato Folk Taxonomy
1976	Clyde Eastman	Rural Sociologist	Ph.D.	USA	Production trends, development
1979-present	Robert Rhoades	Anthropologist	Ph.D.	USA	Post-harvest, farming systems, adoption of varieties
1979-80	Jorge Recharte	Anthropologist	M.A.	Peru	Chanchamayo Valley, Ecology and change
1979, 1983-84	Pierre Bidegaray	Anthropologist	B.A.	Peru	Yurimaguas. Adoption of improved potato varieties
1983	Mario Egoavil	Anthropologist	M.A.	Peru	Storage in a highland community
1984-present	Angelique Haugerud	Anthropologist	Ph.D.	USA	Farming Systems, varieties, Rwanda
1984	Ella Schmidt	Anthropologist	M.A.	Peru	Adoption of improved varieties
1977-present	Marisela Benavides	Sociologist	B.A.	Peru	Production costs (Mantaro Valley), Informal surveys
1979-83	Susan Poats	Anthropologist	Ph.D.	USA	Nutrition, consumption
1984	Norio Yamamoto	Anthropologist	Ph.D.	Japan	Ecology and change
1984	Gordon Prain	Anthropologist	Ph.D.	U.K.	On-farm research, Mantaro Valley

ANNEX 3:

Three Brief Agricultural Anthropological Studies for Interdisciplinary Team Research

Evaluation of Solar Dehydration Techniques

R. Werge

Social Science Department

Introduction

The objective of this report is to compare the black box with traditional methods of solar drying in the Mantaro Valley in Peru. The evaluation consists of four parts: report on a series of experiments, economic evaluation, socio-cultural evaluation and a set of recommendations. (For an illustration of the "black box" see page 30.)

Initial efforts in CIP's investigation of post-harvest technology concentrated upon the development of an improved solar dehydrator which would: (1) reduce the amount of time needed to dehydrate potato products and (2) assure high standards of nutritional and culinary quality. In the early research a "black box," was utilized which was a simple wooden box, painted black, with a removable plastic top. According to the first evaluations, the "black box" showed promise as a means of accomplishing both objectives.

In June and July of 1977, the black box went through a series of tests to measure its effectiveness in drying time compared to indigenous drying methods in the Peruvian Sierra. These experiments were carried out in the town of Concepción, close by CIP's experimental station in Huancayo, with the help of local residents and a rural sociologist. In addition, a short economic and sociocultural study was carried out to determine how the black box or other promising processing technology would be evaluated and accepted by local farmers.

The experiment involved drying of locally processed food products prepared by farmers. Traditionally sun drying is the final stage of the processing of local food products. With the help of local farmers, these products were prepared up until this final stage. The products were then divided into one kilo samples. Samples were placed in the black box, as well as on locally utilized drying surfaces. Locally utilized drying surfaces included cloth sacks, tin sheets, and straw.

During the drying period, the samples were weighed daily to measure water loss. The samples were weighed when the product was judged by the experimenters and local farmers to be fully dried. Local residents also helped to decide if there was a notable difference in the perceived quality of the samples.

59

In addition, to potato products, other locally processed foods made from oca, olluco, maize and barley were dried. These products were included because it was discovered from a study of local technology that drying methods used by farmers were not crop or product specific but rather the same methods were used to dry all of the farmer's products. This meant that any new method would be more acceptable and cheaper to farmers if it were used on the large variety of products dried.

Experimental Results

A total of 20 experiments were carried out over a 4-week period when many local processing activities are normally done. During this period there was some cloud cover but little precipitation.

The results of four experiments which dealt with potato products are presented in Figures 1-4. The results of the other experiments were all similar to these four. In each experiment the black box was tested against one or more traditional drying methods.

The graphs show very little difference in drying rate between the methods tested. The curves show an initial rapid loss in moisture followed by a slower rate. No single method shows a consistent superiority, nor was there a notable difference in quality perceived by local farmers except in the case of a product made from olluco which had a yellower color when dried in the black box.

One of the factors influencing the performance of the black box was the lack of ventilation for drying the product. Moisture trapped inside the black box by the plastic top cut down on the amount of incoming solar radiation. The top was removed in subsequent experiments, but since all indigenous methods expose the product fully to air movement, the open black box did not prove to perform substantially better.

Several additional experiments showed, however, that the size and cut of the product was an important variable in drying time and rate. Smaller sized chunks of potato dried more rapidly than large sized potatoes owing to the greater percentage of area exposed to the sun and wind. If future experiments continue to emphasize drying time and rate, the cutting and preparation of the product should be given more attention.

Economic Evaluation

Table 1 presents a series of cost estimates for using the different materials tested. The table assumes that the drying materials will be used only during the normal period for processing in the region; May through August. In fact, tin sheets and sacks are used for other purposes during the year and it can be assumed that the black box similarly might be used for storage or other purposes. The estimated cost of the black box is S/. 2,000, but periodic replacement of plastic and paint increase the total over its estimated 6-year life period. The cost of straw is based on an estimated one hour of work to collect enough grass to cover one square meter.

The cost figures per week are perhaps not as important in terms of potential adoption patterns as are the initial cost figures. The use of the black box by farmers would assume an initial outlay of S/. 2,200 whereas their currently used

Table 1. Estimated costs of solar drying methods

	Initial cost S/.	Life years	No. of weeks used/year	Cost/week S/.
Tin sheet	100.00	10	16	0.62
Cloth sack	100.00	5	16	1.25
Straw	8.00	1	16	0.50
Black box	2,200.00	6	16	22.92

methods involve a very small initial outlay, at most S/. 100.[1] Since the materials currently used have other functions, however, their cost is hidden. Farmers consider that they make no outlay at all for drying food products but merely use materials which are already available.

Sociocultural Evaluation

The farmers of the Mantaro Valley are innovative in terms of their processing technology. Many households, for example, have purchased small meat grinders over the last decade which are used instead of rough stones for grinding and mashing foods. The milk plant in Concepción is an example of a successful large scale innovation requiring farmers to shift part of their agricultural production into milk production. Farmers' willingness to adopt new technology, however, is vitally dependent upon the costs and benefits of the new technology and other factors.

One important factor in the adoption of new technology by farmers is their view of their current technological problems. In a survey of 124 households in Concepción, farmers cited "low yields" and "lack of manpower" for not engaging in more potato processing. The making of *chuño* and *papa seca* were viewed as laborious, but within each type of processing, different steps presented the major difficulty. In making *papa seca,* for example, the most labor intensive stage was peeling the potatoes after boiling. For no product was sun-drying viewed or mentioned by farmers as a limiting factor. The length of time necessary for drying was not viewed as being unnecessarily long nor was any clear advantage seen in drying the product faster.

Given the fact that drying time does not appear to be an important variable, the adoption of the black box will hinge upon its ability to dry a high quality and notably different product which would command a high market price.

If such a development does occur, then the adoption of the innovation would depend upon the identification of a particular type of person who would be most apt to adopt a new processing innovation. Specialized processors do operate in the Mantaro Valley, principally as operators of mills for grinding grains or small businesses for making *maiz pelado* from maize.[2] These processors occupy a

---

[1]Assuming that the cost of constructing a black box could be halved, it would still represent a cost 10 times greater than the next most expensive method.

[2]In American English, this product is known as *hominy;* it is maize which has been boiled with calcium to remove its skin, and subsequently dried. Two of these maize processing businesses operate near Huancayo. The operators of the *maiz pelado* businesses dry their product in the sun on simple concrete floors.

specialized niche in the local food processing industry and may be potential clients for processing technology rather than farmers who are working principally for their own consumption or cooperative whose need for processing technology appears to be very subject to differing yield and market conditions from year to year.

Summary and Recommendations

This research indicates that the black box technology is limited in its appeal due to: relatively high cost, inability to demonstrate a clear advantage over current methods in terms of drying time, and the farmers' perception that other factors in the processing systems are of more crucial importance.

Based on these findings, future processing experiments should consider the following options:

1. Concentrate on the problem of creating a new or perceivably higher quality product than currently being produced, by using solar drying as only one of the processing stages.
2. Concentrate upon new or adapted technology which could solve problems presented by other phases of the traditional processing system, such as peeling.
3. Concentrate upon the developing shorter drying time capacity in the black box, approaching the problem from a theoretical standpoint with less concern about short-term applicability.

The first option, which would involve continuing research on developing a complete process rather than only on a solar dryer would seem to provide the greatest flexibility for the processing research program and provide the greatest opportunity for the continuous testing of innovations in conjunction with farm-level technology.

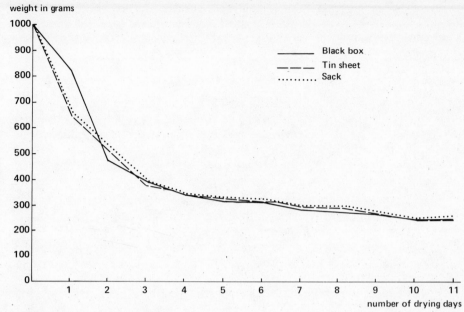

Figure 1.

weight in grams

number of drying days

Black box
Tin sheet
Sack

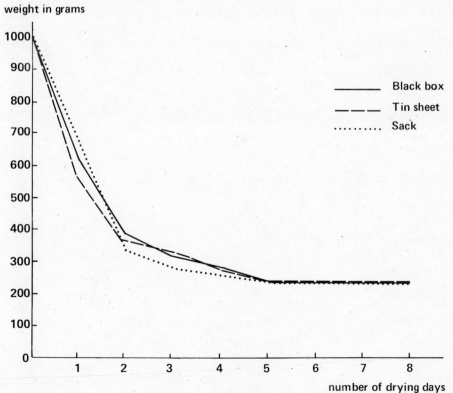

Figure 2   Papa Seca

weight in grams

number of drying days

Black box
Tin sheet
Sack

63

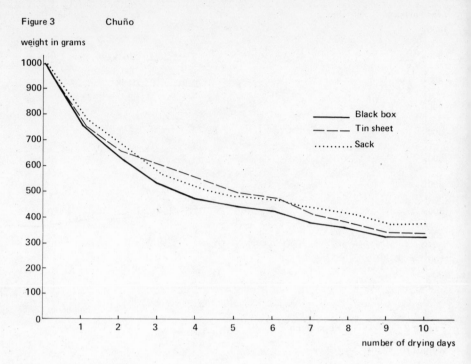

Figure 3   Chuño

weight in grams

Black box
Tin sheet
Sack

number of drying days

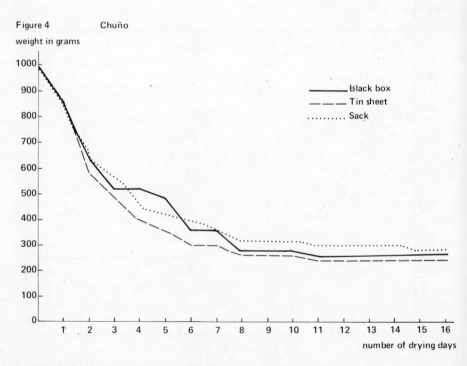

Figure 4   Chuño

weight in grams

Black box
Tin sheet
Sack

number of drying days

Recommendations for On-farm Agro-economic Trials: Cañete, 1980

Robert Rhoades
Marisela Benavides

An informal survey over 10 days in late February and early March, 1980, was to help CIP's agroeconomic team focus better on farm-level problems in conducting trials. The agronomic team had already been through one season of trials and had considerable knowledge of the Cañete Valley. However, they had conducted trials mainly with larger farmers located in the center of the valley and on problems generally defined from the outside rather than by farmers. The attempt then was to (1) define agroecological zones; (2) identify representative types of farmers and their perceptions of production problems; (3) interview local extension and ministry workers to better pinpoint relevant problems for which improved technology might be available. This report, highly simplified, presents the findings of this informal survey.

Agro-ecological Production Zones

We would strongly urge that the upcoming trials take into account the internal agroecological diversity of the valley. Mainly based on soil, irrigation, and socioeconomic conditions, we have determined the existence of three main zones: (1) Upper Valley Margin, (2) Valley Center, and (3) Lower Valley Saline Margin. In each of these zones farmers have distinct sets of production problems or possibilities. Although we cannot present the mass of data we have available, we can briefly summarize the differences in the valley.

Upper Valley Margin

This is a zone of small agriculturists with most holdings varying from 1 to 3 hectares. It contains the poorest soil in the valley, being quite shallow, sandy, and rocky. Since the agriculturists here receive water only every 8 to 10 days (by *mita)* and face water management problems, water supply is considered a major problem. They tend to opt for plants which require less water, mainly cotton (which also has price stability). The water problem has been severe for 2 years because of a lack of rains in the highlands and predictions are that this year will see an even greater scarcity of water. Here one finds the greatest variation in crops and intercropping. Potatoes are grown mainly in the sectors of Quilmana Alto and Nuevo Imperial (47% of all Cañete potato growers for 1980 are programmed in these areas according to Ministry data).

Valley Center

The valley center was historically the location of the large farms which are cooperatives today. This region contains the "alluvial plains" soils, considered to be deep and the best in the valley for agriculture. Scattered throughout the valley are sections of small and medium-size agriculturists. The cooperatives and medium-scale producers concentrate on three or four main commercial crops (cotton, potatoes, maize, sweet potato). They farm with tractors and have a strong market orientation. The small agriculturists in the zone also plant commercial crops but also pursue cultivation for home use *(pan llevar)*. The cooperatives receive water at all times so there are no major water problems. A typical rotation is cotton-potatoes-maize.

Lower Valley Saline Margin

The main characteristics of soil are: salinity, clay texture, and poor drainage. Farmers complain especially of salinity which they note prevents the growing of potatoes. Cotton, maize and pasture for cattle are the main crops here. The zone has a mixture of *CAPs* (cooperatives) and small to medium producers. Under a land rehabilitation project, more than 3,000 hectares will be improved for cultivation purposes. Since most irrigated arid zones have problems with salinity, Cañete could be an area for fruitful investigation with application to other world arid zones. Many farmers in this area wish to plant potatoes but are not willing to take the risk because of salinity.

Types of Farmers and Farmers' Perceptions of Problems

In addition to identification of major zones, we also studied six areas within the valley. It was learned that significant variation occurs in agricultural practices even within our larger zones. Each area has its own special characteristics (demographic patterns, crops, irrigation system) and anyone doing experiments would benefit enormously from the detailed studies on 17 areas of the valley carried out by the agronomists of Valle Grande. This information includes complete and detailed questionnaires, often covering every farmer of the selected area.

Contrary to popular belief, Cañete is not a valley of only large farming operations. It is also a farming community made up of small landholdings. According to 1976 data, 84.2% of all farm units contain less than 3 hectares, 11.2% with 3 to 9.9 hectares and the other units are medium or *CAPs*. It is also not an established fact that mainly medium size farmers and *CAPs* grow potatoes. In fact, according to the Ministry's registration (all farmers must submit a cultivation plan) the average size planting is around 4 hectares. The vast majority plant only 2 to 3 hectares. Furthermore, since 1976 the *CAPs* have drastically cut the number of hectares they plant in potatoes. According to 1980 ministry data (not quite complete for 1980) only 25% of total hectareage will be planted by *CAPs* this year.

The sectors Quilmana Alto, Quilmana Bajo and Nuevo Imperial Alto contain most of the potato farmers (69.75%), and nearly all of those farm less than 3 hectares of potatoes. These three areas account for 48.25% of the programmed land area. Only in Quilmana Bajo do we find *CAPs* and a significant number of medium farmers. In Nuevo Imperial the average size of planting will be 1.73 (N = 99). It should be further noted that 86.74% of all planting will take place in April and May. Thus, if one can speak of an "average" farmer (representative of the majority of the region) it would be a farmer with 2 to 3 hectares who lives in one of the marginal communities and plants in April and May. In any case, these available data suggest that if representativeness is a concern then at least 70% of the experiments should deal with these small farmers.

Farmers' Perceptions of Production Problems

To acquire a better understanding of farmers' perception of problems we conducted a non-random survey with 47 farmers from various zones. Among other things, we asked them to rank their production problems in order of importance.

Farmer Ranking of Production Problems

	No. of farmers	% of total
1. Cost of seed	32	65
2. Cost of inputs (besides seed)	31	63
3. Irrigation problems	19	39
4. Insects	18	37
5. Soils (poor or saline)	13	26
6. Disease	11	22
7. Marketing	11	22
8. Climate	6	12
9. Others	8	16

The farmers of Cañete are presently weighing the decision whether to plant. This may have biased our survey, but there is little doubt that the prevailing cost of seed (105 soles/kilo) is a major concern in the valley. Nearly all farmers mentioned "risk," in conjunction with cost of seed and other inputs. Potatoes are extremely expensive and a crop failure would be a strong financial setback for small farmers.

Each agroecological zone has it own type of production problems. In addition to their concern with costs, the farmers of the marginal zone rank irrigation as a major problem (N = 14, 50% of total farmers in the zone). In the central zone (where they mainly receive water continuously) problems with insects *(mosca minadora)* was a major concern (N = 7, 50% of central zone farmers). Along the coastal-saline zone, farmers perceive soil problems as the next problem after cost of seed. This is due to high salinity, poor drainage, and what they call a lack of *aqua dulce* (sweet water) since they are at the end of the irrigation channel and received the water after it has gone through the entire system. The implications of this data on perception of problems are that seed storage experiments may benefit farmers in all zones (including *CAPs* in the center of the valley), irrigation experiments may be most beneficial on the margins where water problems exist. Of course, experiments with salinity along the coast may be worthwhile if the transferability to other world zones is an objective of the experiments (as stated in the justification of the Cañete project).

Interviews with Extension Workers and Ministry Officials

In addition to our work with farmers we also interviewed extension workers and ministry officials. Eng. Trelles, who is in charge of the potato section at the Ministry, recommended three broad categories of experiments:

1. Nematode Control (cultural practices, chemical control or resistant varieties).

2. Water/Irrigation (any experiment to help solve water problems of small farmers).
3. Salinity (with the opening of 3,000 new hectares, potatoes could play a role if we had the varieties or agronomic techniques to deal with salinity).

Other extension workers added the following: (1) storage experiments; need to store *criolla* seed from September/October to March; (2) insect control, especially *mosca minadora*, (3) fertilizer trials, incorporating more organic materials in soil or trials with *guano de corral*, (4) biological control of insects.

Conclusions

Based on all the evidence we have available we would recommend the following:

1. Most of the experiments (70%) should be carried out with small farmers living in the marginal zone, the remaining 30% among medium size and *CAPs*.
2. The experiments should aim to increase efficiency in seed use, decrease cost of seed and inputs while increasing output (high cost complex packages will only increase the risk factor).
3. The key problems identified by farmers should be addressed: "mosca minadora," water problems, cost of seed, and salinity (in one zone).
4. Most farmers did not identify nematodes or storage as problems but these were stressed by knowledgeable extension workers in the area.
5. The experiments should be relevant to the agroecological zone where the experiment is conducted and to the majority of farmers in the zone.
6. Traditional fertilizer trials should be pursued only after an extensive review of the data on current farmer fertilizer practices available in Valle Grande, a local private research institute.

Mixed Cropping Practices in the
Northern Peruvian Andean Sierra

Robert Werge
Social Science Department

1976

Problem

The potential for mixed cropping of potatoes with other crops in the Peruvian Sierra has been frequently discussed at the Center. In order to examine current practices in rural areas, a quick survey of some 275 *chacras* (fields) was made in the Callejon de Conchucos (Ancash) on 17 April 1976. The *chacras* were found along a trail leading from the town of San Marcos at 2,964 meters to the community of Pujun at 3,800 meters. These are the results of that survey:

1. *The most common practice followed was monoculture, i.e., the planting of a single crop in one field.* Eighty-six percent of the fields counted were planted in this manner. Among the crops planted were:

Table 1

Crop	No. of Fields	% of total Chacras in monoculture
Trigo (wheat)	95	41%
Papa (potatoes)	76	33%
Maiz (maize)	32	14%
Cebada (barley)	9	4%
Arveja (pea)	7	3%
Tarhui (Andean grain)	5	2%
Others	8	3%
TOTAL	232	100%

2. *Only 14% of the chacras were planted in more than one crop.* Of these 40 fields, 29 had intermixed crops, either planted in adjacent rows or in more scattered fashion. The various combinations found were:

Table 2

Main Crop	2nd Crop	No. of Chacras
Maiz	frijol	16
Maiz	quinua	7
Maiz	papa, cebada	1
Maiz	frijol, calabaza	1
Maiz, frijol	quinua, calabaza	1
Maiz, papa	arveja, haba, calabaza	1
Quinua	haba, tarhui	1

with few exceptions, *maiz* was the dominant crop with the second and subsequent crops planted with much less frequency.

3. *In both cases where "papa" (potatoes) was intermixed, it was planted in alternate rows with "maiz."* In the first case, "arvejas" were planted in the same rows as *papa* while *arveja* and *habas* were planted with the *maiz. Calabazas* were planted randomly. This particular field was a small *chacra* planted on an especially eroded and poor soil.

The second case in which *papa* was intermixed was an example of *sequential* planting based on alternate rows of *maiz* and *papa*. As the *papa* was harvested, *cebada* was planted in its stead.

4. There were 11 *chacras,* all above 3,300 meters, in which strip planting was cultivated. This system is quite distinct in appearance and consists of cross-cutting rows of *quinua* which create square plots in which *oca* or *olluco* are grown. In one example, *mashua* was grown with the *oca* and *olluco*. The following diagram describes this pattern.

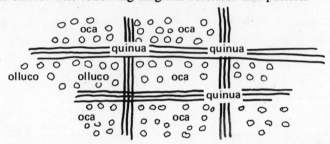

Farmers stated several times that this pattern is also employed with potatoes as one of the tubers, but no such fields were seen.

5. *Among the most planted crops, the pattern of interplanting varies considerably.* Wheat, for example, is always planted alone, while *quinua* is always planted with other crops.

Table 3

| | | % of fields | | |
| | | | Mixed cropping | |
Crop	No. of fields	monocrop	with 1 crop	with 2/+ crops
Trigo			--	--
Papa	78	97	--	3
Maiz	59	54	38	8
Quinua	20	--	43	57

while not included in this sample, household gardens *(huertos),* may represent more intensive interplanting than the *chacra*. It would be interesting to see how potatoes are planted in the household garden.

Conclusion

1. From this admittedly small sample, it appears that mixed cropping in the north Sierra is of a very limited variety. The overall tendency is toward monoculture. Variety exists, however, in the sense that each farmer has a number of small fields, scattered at various altitudes, each of which contains an example of the basic crop complex (*papa, maiz, trigo,* etc.).

2. Where mixed cropping does exist, the overall tendency is to intermix legumes with *maiz* or to strip plant *quinua* and the indigenous tubers. The first pattern is commonly encountered at the lower altitudes and the second much higher up. The distinctiveness of the strip planting system is, at this point, a bit difficult to explain.

3. Potatoes are almost never associated with other crops. In 76 of the 78 *chacras* in which they were found, potatoes were grown alone. It remains to be seen if this percentage holds true in other parts of the Sierra.

4. It would appear that while mixed cropping is practiced, the overwhelming preference is to plant crops, such as potatoes and wheat, by themselves. This may be due to climatic/ecological reasons, to the relatively small number of possible crops available, to mechanical (harvesting, planting) incompatibilities, or to other cultural practices. At the same time, account should be given of the fact that a crop such as guinua is never found growing by itself, but only in association with other crops.

Future studies of mixed cropping practices in Peru, especially as they involve or do not involve potatoes, should consider:

1. Actual farmer practices in a variety of regions, and
2. The peasant "theory" behind the preference for monoculture and the limited mixed cropping patterns which exist.

Such studies could help to guide and design our own experiments — with mixed cropping and new potato varieties.